Food Culture and Politics in the Baltic States

This book focuses on food culture and politics in three Baltic states: Estonia, Latvia, and Lithuania. In popular and scholarly writings, the Baltic states are often seen as a meat-and-potatoes kind of place, inferior to sophisticated cuisines of the West and exotic diets in the East. Such views stem from the long intellectual tradition that focuses on political and cultural centers as sources of progress. But, as a new generation of writers has argued, in order to fully grasp the ongoing cultural and political changes, we need to shift the focus from capital cities such as Paris, Berlin, Rome, or Moscow to everyday life in borderland regions that are primary arenas where such transformations unfold. Building on this perspective, chapters featured in this book examine how identities were negotiated through the implementation of new food laws, how tastes were reinvented during imperial encounters, and how ethnic and class boundaries were both maintained and transgressed in Baltic kitchens over the course of the twentieth and early twenty-first centuries. In so doing, the book not only explores culinary practices across the region, but also offers a new vantage point for understanding everyday life and the entanglement between nature and culture in modern Europe. This book was originally published as a special issue of the *Journal of Baltic Studies*.

Diana Mincytė is an Assistant Professor of Sociology in New York City College of Technology of the City University of New York. The recipient of numerous grants and fellowships, she publishes on social and environmental dimensions of agro-food systems both in and outside of post-socialist East Europe.

Ulrike Plath is a Professor of German culture and history in the Baltic region at Tallinn University and a senior researcher at the Under and Tuglas Literature Centre of the Estonian Academy of Sciences. She has published in the areas of Baltic cultural history of the Enlightenment and Baltic foods, gardening, and environmental history.

Food Culture and Politics in the Baltic States

Edited by
Diana Mincytė and Ulrike Plath

LONDON AND NEW YORK

First published 2017
by Routledge

2 Park Square, Milton Park, Abingdon, Oxfordshire OX14 4RN
52 Vanderbilt Avenue, New York, NY 10017

Routledge is an imprint of the Taylor & Francis Group, an informa business

First issued in paperback 2020

British Library Cataloguing in Publication Data
A catalogue record for this book is available from the British Library

ISBN 13: 978-1-138-70293-6 (hbk)
ISBN 13: 978-0-367-59512-8 (pbk)

Typeset in Perpetua
by diacriTech, Chennai

Publisher's Note
The publisher accepts responsibility for any inconsistencies that may have arisen during the conversion of this book from journal articles to book chapters, namely the possible inclusion of journal terminology.

Disclaimer
Every effort has been made to contact copyright holders for their permission to reprint material in this book. The publishers would be grateful to hear from any copyright holder who is not here acknowledged and will undertake to rectify any errors or omissions in future editions of this book.

Contents

Citation Information

The chapters in this book were originally published in the *Journal of Baltic Studies*, volume 46, issue 3 (September 2015). When citing this material, please use the original page numbering for each article, as follows:

Chapter 1
Introduction: Exploring Modern Foodways: History, Nature, and Culture in the Baltic States
Diana Mincytė and Ulrike Plath
Journal of Baltic Studies, volume 46, issue 3 (September 2015) pp. 275–281

Chapter 2
Good, Clean, Fair … and Illegal: Paradoxes of Food Ethics in Post-Socialist Latvia
Guntra A. Aistara
Journal of Baltic Studies, volume 46, issue 3 (September 2015) pp. 283–298

Chapter 3
Geographies of Reconnection at the Marketplace
Renata Blumberg
Journal of Baltic Studies, volume 46, issue 3 (September 2015) pp. 299–318

Chapter 4
Changing Values of Wild Berries in Estonian Households: Recollections from an Ethnographic Archive
Ester Bardone and Piret Pungas-Kohv
Journal of Baltic Studies, volume 46, issue 3 (September 2015) pp. 319–336

Chapter 5
"Is that Hunger Haunting the Stove?"
Leena Kurvet-Käosaar
Journal of Baltic Studies, volume 46, issue 3 (September 2015) pp. 337–353

Chapter 6
The Evolution of Household Foodscapes over Two Decades of Transition in Latvia
Lani Trenouth and Talis Tisenkopfs
Journal of Baltic Studies, volume 46, issue 3 (September 2015) pp. 355–375

Chapter 7
The Making of the Consumer? Risk and Consumption in Europeanized Lithuania
Ida Harboe Knudsen
Journal of Baltic Studies, volume 46, issue 3 (September 2015) pp. 377–391

Chapter 8
Atlantic Herring in Estonia: In the Transverse Waves of International Economy and National Ideology
Kadri Tüür and Karl Stern
Journal of Baltic Studies, volume 46, issue 3 (September 2015) pp. 393–408

Notes on Contributors

Guntra A. Aistara holds a PhD from the University of Michigan's School of Natural Resources and Environment. She is an environmental anthropologist, whose research interests include organic agriculture movements, agrobiodiversity and seed sovereignty, the political ecology of small farmers' struggles over the control of land and seeds in the face of free-trade agreements, and socioecological resilience of local food systems. She is an Assistant Professor in the Department of Environmental Sciences and Policy at the Central European University in Budapest, and was an Agrarian Studies fellow at Yale University from 2014 to 2015.

Ester Bardone is a Lecturer in Ethnology at the Institute for Cultural Research and Fine Arts, University of Tartu, where she received a PhD in Ethnology. Her research interests include rural tourism, small-scale rural entrepreneurship, and heritage production in Estonia, and changes in Estonian food culture.

Renata Blumberg received her PhD in Geography from the University of Minnesota in June 2014. She is currently an Assistant Professor in the Department of Nutrition and Food Studies at Montclair State University. Thus far, she has studied alternative food networks in Latvia, Lithuania and the United States, and has paid particular attention to their sociospatial dimensions.

Ida Harboe Knudsen is a postdoctoral fellow at the University of Aarhus, Denmark, a position financed by the Danish Independent Research Council, Culture and Communication (Det Frie Forskningsråd, Kultur og Kommunikation [FKK]). She obtained her PhD from the Max Planck Institute for Social Anthropology in Halle, Germany, in 2010, and has written on topics including agriculture, EU enlargement, and dairy production, as well as on illegal work and mafia affiliations in Lithuania.

Leena Kurvet-Käosaar is an Associate Professor of literary theory at the University of Tartu and a senior researcher of the Archives of Cultural History of the Estonian Literary Museum. She is the author of *Embodied Subjectivity in the Diaries of Virginia Woolf, Aino Kallas and Anaïs Nin* (2006) and the editor (with Lea Rojola) of *Aino Kallas, Negotiations with Modernity* (2011), a special issue of *Methis* on Estonian life writing (2010) and numerous articles of Baltic women's deportation stories within the framework of critical trauma studies.

Diana Mincytė is an Assistant Professor of Sociology in New York City College of Technology of the City University of New York. The recipient of numerous grants and fellowships, she publishes on social and environmental dimensions of agro-food systems both in and outside of postsocialist East Europe.

Ulrike Plath is a Professor of German culture and history in the Baltic region at Tallinn University and a senior researcher at the Under and Tuglas Literature Centre of the Estonian Academy of Sciences. She has published in the areas of Baltic cultural history of the Enlightenment and Baltic foods, gardening, and environmental history.

Piret Pungas-Kohv is an expert in the area of environmental awareness at the Estonian Nature Fund where she focuses on a large-scale wetland restoration project. Since completing her PhD in human geography at the University of Tartu, she has studied place-making practices in various natural habitats and diverging meanings of specific places, especially mires as significant loci in Estonian culture.

Karl Stern is a PhD student in the Department of Contemporary History at the University of Tartu. His research focuses on the application of nontariff measures in the protectionist trade policies of the Republic of Estonia in the 1930s. He is working in the Internal Market Department of the Ministry of Economic Affairs and Communications.

Talis Tisenkopfs is a Professor of Sociology at the University of Latvia, Faculty of Social Sciences, and Director of the Baltic Studies Centre. He is the Vice-President of the European Society for Rural Sociology and is the author of scientific articles and literary sociological essays. His current interests include sustainable food systems, knowledge and innovation processes in agriculture and rural development, and post-socialist transformation of agriculture and rural areas. He has also led a number of research projects underwritten by the European Union funding agencies.

Lani Trenouth is a Marie Curie Early Stage Research Fellow in the PUREFOOD project 2011–2014. She is a doctoral candidate at the Wageningen University and is based at the University of Latvia researching alterity in food systems and food consumption practices using visual research methods in Latvia.

Kadri Tüür is a PhD student in the Department of Semiotics at the University of Tartu. Her research focuses on Estonian nature writing and its analysis. She has published articles in the areas of ecocriticism and semiotics, co-edited several article collections, and co-organized conferences in the fields of semiotics and literary studies.

Acknowledgments

The first conversation that led us to put together the special issue on food culture in the Baltic states in the *Journal of Baltic Studies*—which became this book—took place at the summer school for environmental history on an island off the coast of Estonia in the spring of 2011. The summer school, and our subsequent meetings, were supported generously by the Rachel Carson Center for Environment and Society in Munich, Germany, for which we are deeply grateful. The Center, as we came to call it, has left a lasting impact on this book not only because it provided institutional and financial support, but also because it opened a stimulating intellectual space to think about food as a political, social, and environmental practice.

In the process of editing the special issue, we met and worked with a number of people whose encouragement and unwavering support made its publication possible. Bradley Woodworth introduced us to the idea of representing all three Baltic states and encouraged us to pursue the publication. At the *Journal of Baltic Studies*, we appreciated the opportunity to work with editor Terry D. Clark and his team whose constructive feedback, editorial support, and enthusiasm guided us at critical moments. At Routledge Press, we thank series editor Emily Ross and her team for accommodating our requests and needs when transforming the special issue into a book.

We are grateful to reviewers who prepared carefully calibrated comments for each author. Without such feedback, the papers would not have been as insightful and lucid as they have become. We thank Paulius Narbutas and Robert M. Wills for their help with language editing in efforts to create a more coherent style across the issue.

Most importantly, we thank the authors who entrusted us with their work and their patience with a lengthy revision process. Their hard work and persistence forms the foundation of this book. We hope that the readers will be able to appreciate the confluence of ideas that resulted in this book and multiple dimensions of food politics and culture woven into it.

INTRODUCTION

EXPLORING MODERN FOODWAYS: HISTORY, NATURE, AND CULTURE IN THE BALTIC STATES

Diana Mincytė and Ulrike Plath

In her article about food practices in Italy, Alison Leitch (2003) argues that the burgeoning interest in gastronomy has become a medium through which the public expresses deeply seated concerns with the shifting political, economic, and industrial order in Europe. If Leitch is correct about the importance of food culture, then what do culinary practices in the Baltic States tell us about life, culture, identities, and historical memory in the region? How can a closer look at kitchens, tastes, and changing shopping habits in three Baltic nations illuminate experiences and practices of living, eating, and working in this part of the world today and historically?

The purpose of this special issue is to explore the above questions by looking at the changing foodways in Estonia, Latvia, and Lithuania in the twentieth and at the beginning of the twenty-first centuries. Selected from a large pool of proposals, the articles featured in this issue examine the ways in which ethnic, national, and class boundaries were both maintained and transgressed through diets, how agro-food systems were transformed through imperial encounters and national regimes, and how identities were reinvented through food procurement at different historical junctures. At the same time, the authors have been careful not to reduce food to a mere lens through which to study social history, politics, and culture (Gille 2009). For them, food in-and-of itself is an embodied practice and lived experience, constituting a material link between the human body and the environment. Following this approach, this special issue speaks directly to the growing interdisciplinary body of scholarship that places food in larger environmental and historical cycles, taking seriously Donald Worster's maxim that environmental history goes through one's belly (Mink 2009).

This collection of articles builds on and contributes to the long scholarly tradition of studying food and everyday life in the region. Much like in the rest of the European academy, such studies have tended to follow four distinct disciplinary tracks, including ethnological tradition, historical research, literary studies, and a combination of social science approaches, including anthropology, geography, political science, and sociology.

Ethnologists and folklorists were first in their efforts to systematically collect, document, and analyze local food practices since the early twentieth century. In the spirit of national movements, they were primarily concerned with preserving what they considered as the relics of early national traditions of the peasants, turning a blind eye on modernization of food culture in and around multicultural towns. Despite these limitations, it is thanks to these ethnological expeditions that we are left with a plethora of artifacts, drawings, photographs, voice recordings, and copious notes providing a glimpse into the everyday life of rural households (Dundulienė 1963; Moora 2007; Troska and Viires 2008). Today, scholars continue working with these archives in efforts to draw new connections and explore topics such as changing berry-picking conventions in modern Estonia (Bardone and Pungas-Kohv, this issue).

Unlike ethnologists and folklorists, historians have been more concerned with how food cultures of all classes relate to cultural, economic, technological, and climatic changes, and particular historical events (Plath 2012). They have examined such topics as, for example, food assistance during the bouts of famine in the seventeenth through nineteenth centuries (Liiv 1938; Seppel 2008), alcohol production and consumption practices in the context of industrialization in the region (Astrauskas 2008), and the importance of food substitutes in the culinary networks tying the Baltics and the rest of the world (Plath 2008). In this context, maritime endeavors have been of particular significance because they opened the gates to faster cultural exchanges, as shown by the historians of Medieval Livonia (Põltsam-Jürjo 2012, 2013; Mänd 2012; Sillasoo 2013) and those studying the twentieth century (Tüür and Stern, this issue).

More recently, a new generation of literary scholars have started to publish textual and narrative analyses of fiction, memoirs, and other literary texts on food culture, combining ethnological, historical, and semiotic approaches under the umbrella of cultural studies. Such analyses include studies of the printed cookbooks (Dumpe 1998; Viires 1985), studies of food terminology (Mincytė 2011; Paškevica 2012), interpretative and discursive analyses of literary works of art (Plath 2013; Ross 2012), and the textual examination of the popular press (Tüür and Stern, this issue). Among literary scholars, there is also a growing interest in studying the role of food in biographies, memoirs, and life stories (Kurvet-Käosaar, this issue).

The fourth line of inquiry combines a swath of studies that have drawn on ethnographic field methods to examine Baltic food practices in the backdrop of ongoing globalization, liberalization, and austerity politics. Often relying on Marxian critiques of capitalism, this scholarship is anchored in the debates about food production, consumption, and distribution cycles (McMichael 2009; DuPuis and Goodman 2005) and has engaged in the burgeoning discussions about post-socialism. Examples include inquiries into the formation of consumer society in eastern Europe (Caldwell 2011; Vonderau 2010; Knudsen Harboe, this issue; Trenouth and Tisenkopfs, this issue), experiences of injustice, power and disempowerment (Dunn 2005; Gille 2011; Klumbytė 2009; Aistara, this issue), national identity construction and reproduction (Lankauskas 2002; Võsu and Kannike 2011), and the social and material embeddedness of food practices in the historical and spatial junctions (Blumberg, this issue; Aistara 2014).

Despite the shared interest in food-related topics, however, there are surprisingly little exchanges spanning the four research traditions surveyed above. This might be due to the institutionalization of disciplinary boundaries that make it difficult to develop sustained intellectual conversations, but also because there have been few venues for the scholars from the disparate fields to build productive collaborative relations. In this sense, this special issue is one of the first attempts to showcase the range and depth of Baltic food studies and create a space for interdisciplinary exchanges.

The diverse disciplinary backgrounds of contributors to this special issue address these goals. Ester Bardone draws on the ethnological approaches, while Leena Kurvet-Käosaar contributes a literary perspective. Karl Stern is an economic historian who collaborates with ecosemiotician Kadri Tüür. Two other authors – Renata Blumberg and Piret Pungas-Kohv – are geographers. Ida Harboe Knudsen is trained in anthropology, while Guntra Aistara, also an anthropologist, brings a strong grounding in environmental studies. Sociologists Tālis Tisenkopfs and Lani Trenouth are already familiar names in the sociology of food and agriculture in Europe. In each case, the authors have worked to transcend the bounds of their disciplinary backgrounds by engaging the works from other fields in rethinking the role of food in their Baltic history, culture, and daily life.

Contributions and Shared Themes

The special issue opens with two articles that link contemporary food production and consumption experiences in the Baltic States with the history and memory of socialism. Focusing on small-scale, non-industrial food economies, both papers show that the socialist past casts a long shadow over how people in the Baltics construct meanings of food justice, locavorism, and ethical eating today. The first article by Guntra Aistara questions the assumptions behind what constitutes "good, clean, and fair" food in alternative food economies embodied in such initiatives as the Slow Food movement and Fair Trade networks. Aistara argues that because these values have been coded in transnational certifications, they have come to contradict and even undermine the moral orientation, rural livelihood, and economic realities in post-socialist Latvia. Similarly, focusing on three markets in Vilnius, Renata Blumberg's analysis contests common assumptions that farmers' markets operate as sites for reconnecting and reimbedding food in local communities. Relying on ethnographic fieldwork, Blumberg sees farmers' markets as part of a larger retail sector shaped by the peculiarities of the post-socialist experience.

The following two articles take us back to the tumultuous history of the twentieth century to illuminate the continuities and changes in food culture in the region. Rereading ethnological surveys conducted under the purview of the Estonian National Museum, Ester Bardone and Piret Pungas-Kohv examine how the values surrounding wild berry picking and consumption changed in Estonian households over the century. They track major transformations from the restricted berry consumption by the peasants under the rule of Baltic German landlords and the Russian Empire, to berries becoming an export commodity in the first part of the twentieth century, to collective picking excursions under socialism, to picking as a leisure activity today. The second article by Leena Kurvet-Käosaar tracks life in forced labor camps in Siberia after the Second World War to remind us that famine is about a social disturbance as much as an experience with suffering. Examining the memoirs and dairies penned by Baltic women, Kurvet-Käosaar shows how food (or the lack of) during deportation constituted a particular site for redefining women's identities, the domestic order, and the boundaries of community and the nation.

In contrast to the experiences of extreme hunger, the next two articles in the issue grapple with the emerging consumer culture of the late twentieth and the early twenty-first centuries that is epitomized in the seeming abundance of food in supermarkets, markets, and specialty shops dotting modern urban landscapes. Following a similar life-story approach as Kurvet-Käosaar, Lani Trenouth and Tālis Tisenkopfs take a broad view of the Latvian food topography to study changing habits over two decades of post-socialist transformations. In their analysis, Trenouth and Tisenkopfs develop four consumer profiles – the urban professional, the rural farmer, the urban retired, and the urban worker – to distill the most prevalent features of the Latvian consumer culture and to show how differently these groups have experienced and "tasted" post-socialism. In the following article, Ida Harboe Knudsen examines the growing popularity of the "authentic" Lithuanian cuisine that valorizes "healthy" countryside products, defying western nutritional advise such as low-fat meat and dairy products. Harboe Knudsen argues that such a reinvention of the Lithuanian gastronomic taste signals the ambivalence with which Lithuania's residents relate to the ongoing "Europeanization" of the region.

The last paper in the special issue knits together the thematic threads developed in the earlier papers by linking particular consumer tastes with the quest for economic sovereignty and identity politics in the first part of the twentieth century. Focusing on Atlantic herring expeditions organized and sponsored by the Estonian government in the 1930s, Kadri Tüür and Karl Stern show that the interwar government was deeply engaged in European food trade tariff politics and actively, even if unsuccessfully, sought to establish Estonia as a self-sustaining fishing nation. Through an investigation of popular accounts about these expeditions, Tüür's and Stern's analysis provides insights into the struggles over food sovereignty that are surprisingly similar to those being fought today by the avant-garde of the local food movements such as La Vía Campesina (Edelman 2014).

As these summaries indicate, issues surrounding nationality, subjectivity, and modernity constitute key themes that link articles in the special issue. It is therefore unfortunate that in the process of compiling the articles, we were unable to include outstanding papers that would have spoken more explicitly to the issues of multiculturalism and transnationalism that have defined the Baltic region for centuries. We do hope, however, that conversations forged in our exchanges with all the authors will prove to be productive and lead to new collaborations.

Acknowledgments

We thank the authors who entrusted us with their work and their patience with a lengthy revision process as well as the numerous reviewers who prepared carefully calibrated comments for each author. Without such feedback, the papers would have not been as insightful and lucid as they have become. We gratefully acknowledge the contribution of Paulius Narbutas and Robert M. Wills who helped us with language editing in efforts to create a more coherent style across the issue.

Disclosure statement

No potential conflict of interest was reported by the authors.

Funding

We would like to thank the Estonian Science Foundation that provided partial funding for this special issue under the project "History of Baltic Food Culture: Production, Consumption and Culture in the Light of Environmental History" (ETF 9419).

References

Aistara, G. A. This issue. "Good, Clean, Fair… and Illegal: Paradoxes of Food Ethics in Post-Socialist Latvia." *Journal of Baltic Studies* 46 (3): 283–298. doi:10.1080/01629778.2015.1073915.

Aistara, G. A. 2014. "Latvia's Tomato Rebellion: Nested Environmental Justice and Returning Eco-Sociality in the Post-Socialist EU Countryside." *Journal of Baltic Studies* 45 (1): 105–130. doi:10.1080/01629778.2013.836831.

Astrauskas, A. 2008. *Per Barzdą Varvėjo: Svaigiųjų Gėrimų Istorija Lietuvoje*. Vilnius: Baltos Lankos.

Bardone, E., and P. Pungas-Kohv. This issue. "Changing Values of Wild Berries in Estonian Households: Recollections from an Ethnographic Archive." *Journal of Baltic Studies* 46 (3): 319–336. doi:10.1080/01629778.2015.1073916.

Blumberg, R. This issue. "Geographies of Reconnection at the Marketplace." *Journal of Baltic Studies* 46 (3): 299–318. doi:10.1080/01629778.2015.1073917.

Caldwell, M. L. 2011. *Dacha Idylls: Living Organically in Russia's Countryside*. Berkeley: University of California Press.

Dumpe, L. 1998. "Aus der Baltischen Kochgeschichte. Die Ersten Lettischen Kochbücher." In *Kultuuri Mõista Püüdes*, edited by T. Anepajo, and A. Jürgenson, 233–245. Tallinn: Teaduste Akadeemia Kirjastus, Ajaloo Instituut.

Dundulienė, P. 1963. *Žemdirbystė Lietuvoje: Nuo Seniausių Laikų iki 1917*. Vilnius: Valstybinė Politinės ir Mokslinės Literatūros Leidykla.

Dunn, E. C. 2005. "Standards and Person-making in East Central Europe." In *Global Assemblages: Technology, Politics, and Ethics as Anthropological Problems*, edited by A. Ong, and S. J. Collier, 173–194. Malden: Blackwell Publishing.

DuPuis, E. M., and D. Goodman. 2005. "Should We Go 'Home' to Eat?: Toward a Reflexive Politics of Localism." *Journal of Rural Studies* 21 (3): 359–371. doi:10.1016/j.jrurstud.2005.05.011.

Edelman, M. 2014. "Food Sovereignty: Forgotten Genealogies and Future Regulatory Challenges." *The Journal of Peasant Studies* 41: 959–978. doi:10.1080/03066150.2013.876998.

Gille, Z. 2009. "From Nature as Proxy to Nature as Actor." *Slavic Review* 68 (1): 1–9.

Gille, Z. 2011. "The Hungarian Foie Gras Boycott: Struggles for Moral Sovereignty in Postsocialist Europe." *Eastern European Politics and Societies* 25: 114–128. doi:10.1177/0888325410374090.

Klumbytė, N. 2009. "The Geopolitics of Taste: The 'Euro' and 'Soviet' Sausage Industries in Lithuania." In *Food & Everyday Life in the Postsocialist World*, edited by M. L. Caldwell, 130–153. Bloomington: Indiana University Press.

Knudsen Harboe, I. This issue. "The Making of the Consumer? Risk and Consumption in EUropeanized Lithuania." *Journal of Baltic Studies* 46 (3): 377–391. doi:10.1080/01629778.2015.1073925.

Kurvet-Käosaar, L. This issue. "'Is That Hunger Haunting the Stove?' Thematization of Food in the Deportation Narratives of Baltic Women." *Journal of Baltic Studies* 46 (3): 337–353. doi:10.1080/01629778.2015.1073951.

Lankauskas, G. 2002. "On 'Modern' Christians, Consumption, and the Value of National Identity in Post-Soviet Lithuania." *Ethnos* 67 (3): 320–344. doi:10.1080/0014184022000031.

Leitch, A. 2003. "Slow Food and the Politics of Pork Fat: Italian Food and European Identity." *Ethnos* 68 (4): 437–462. doi:10.1080/0014184032000160514.

Liiv, O. 1938. *Suur Näljaaeg Eestis 1695-1697. Lisas: Valimik Dokumente Suurest Näljaajast.* Tartu, Tallinn: Loodus.

Mänd, A. 2012. *Pidustused Keskaegse Liivimaa Linnades 1350-1550*. Tallinn: Eesti Keele Sihtasutus.

McMichael, P. 2009. "A Food Regime Genealogy." *The Journal of Peasant Studies* 36 (1): 139–169. doi:10.1080/03066150902820354.

Mincytė, D. 2011. "Unusual Ingredients: Gastronationalism, Globalization, Technology and Zeppelins in East European Imagination." *Anthropology of East European Review* 29 (2): 1–21.

Mink, N. 2009. "It Begins in the Belly." *Environmental History* 14: 312–322.

Moora, A. 2007. *Eesti Talurahva Vanem Toit*. Tartu: Ilmamaa.

Paškevica, B., ed. 2012. *Tulkojums Ar Grašu. Ēdiena Volodnieciskie Un Starpkultūru Aspekti*. Valmiera: Vidzemes Augstskola.

Plath, U. 2008. "Surrogates in Baltic Nutrition and Culture in the 19th Century." In *The End of Autonomy? Studies in Estonian Culture*, edited by C. Hasselblatt, 52–68. Maastricht: Shaker.

Plath, U. 2012. "Baltic Asparagus: Transnational Perspectives on Gardening and Food Culture (17th-19th Cc.)." In *Tulkojums Ar Garšu. Ēdiena Valodnieciskie un Starpkultūru Aspekti*, edited by B. Paškevica, 40–64. Valmiera: Vidzemes Augstskola.

Plath, U. 2013. "Näljast Ja Näljapsühholoogiast A. H. Tammsaare Romaanis 'Ma Armastasin Sakslast'." In *Armastus Ja Sotsioloogia: A. H. Tammsaare romaan 'Ma Armastasin Sakslast'*, edited by M. Hinrikus, and J. Undusk, 50–71. Tallinn: Underi ja Tuglase Kirjanduskeskus.

Põltsam-Jürjo, I 2012. "'Hääleib', 'Saajaleib', 'Iseleib' – Eesti Leivakultuurist 13.-16. Sajandil." *Tuna. Ajalookultuuri Ajakiri* 4: 14–28.

Põltsam-Jürjo, I. 2013. *Pidusöögist Näljahädani. Söömine-Joomine Keskaja Tallinnas*. Tallinn: Eesti Ekspressi Kirjastus.

Ross, J. 2012. "Naisterahva Mõnitus. Toidu Valmistamine Eesti Nõukogude Olmekirjanduses." *Looming* 1: 83–96.

Seppel, M. 2008. *Näljaabi Liivi- ja Eestimaal 17. Sajandist 19. Sajandi Alguseni*. Tartu: Tartu Ülikooli Kirjastus.

Sillasoo, Ü. 2013. "A Cultural History of Food Consumption in Medieval Livonian Towns." In *Landscapes and Societies in Medieval Europe East of the Elbe Interactions between Environmental Settings and Cultural Transformations*, edited by S. Rossignol, S. Kleingärtner, T. Newfield, and D. Wehner, 316–328. Toronto: Pontifical Institute of Mediaeval Studies.

Trenouth, L., and T. Tisenkopfs, This issue. "The Evolution of Household Foodscapes Over Two Decades of Transition in Latvia." *Journal of Baltic Studies* 46 (3): 355–375. doi:10.1080/01629778.2015.1073927.

Troska, G., and A. Viires. 2008. "Söögid ja Joogid." In *Eesti Rahvakultuur*, edited by A. Viires, and E. Vunder, 264–278. Tallinn: Eesti Entsüklopeediakirjastus.

Tüür, K., and K. Stern. This issue. "Atlantic Herring in Estonia: In the Transversal Waves of International Economy and National Ideology." *Journal of Baltic Studies* 46 (3): 393–408. doi:10.1080/01629778.2015.1073928.

Viires, A. 1985. "Kokaraamatud Kultuuriloo Kajastajatena." *Keel ja Kirjandus* 3: 158–166.

Vonderau, A. 2010. *Leben im Neuen Europa: Konsum, Lebensstile und Körpertechniken im Postsozialismus*. Bielefeld: Transcript.

Võsu, E., and A. Kannike. 2011. "My Home Is My Stage: Restaurant Experiences in Two Estonian Lifestyle Enterprises." *Journal of Ethnology and Folkloristics* 5 (2): 19–47.

GOOD, CLEAN, FAIR ... AND ILLEGAL: PARADOXES OF FOOD ETHICS IN POST-SOCIALIST LATVIA

Guntra A. Aistara

If the Soviet Union perpetuated an economy of scarcity, the European Union maintains an economy of purity: in Soviet Latvia a lack of raw materials restricted production, while in the EU, hygiene regulations restrict processing and sale of homemade foods. In both periods, producers and consumers have cultivated informal social networks that challenge relations to structures of power, equating illegally obtained food products with an ethical stance. Positioning local informal networks as illegal obscures persistent inequalities in access to markets for the smallest home producers, and stigmatizes local practices and social networks as backwards without addressing the causes.

Throughout the world, attention to ethical choices in food consumption is growing, as a reaction to the increasing awareness of the social and environmental consequences of neoliberal policies, processes of economic globalization, and the industrialization of agriculture. This means that consumers exhibit a "growing preference for objects that are produced in ways that are seen to be socially and environmentally good; or at least, better than the alternatives on offer" (Carrier and Luetchford 2012, 2). These trends in ethical food consumption rely on certifications and other ways to demonstrate environmentally friendly production practices (organic), fair wages and working conditions for farmers and workers (fair trade), short supply chains (local), and the humane treatment of animals (vegan, vegetarian, and animal rights movements) (see for example Carrier and Luetchford 2012; Guthman 2004; Moberg 2010 on various types of ethical consumption). Indeed, numerous initiatives attempt to combine this multiplicity of values, as summed up with the slogan of the international Slow Food movement: "good, clean, and fair."[1]

While it is easy to presume that such ethical consumption paradigms offer clear-cut prescriptions of action, concepts such as good, clean, and fair are context-dependent and their meanings shift in relation to political regimes and social circumstances. In this article, I explore specifically how discourses of good, clean, and fair come into uncomfortable dialogue in the sociopolitical context of post-socialist Latvia as a new member state of the European Union. I will trace the transformation and persistence of informal exchange networks for the procurement of home-produced food items in the face of changing power relations from the Soviet Union to the European Union (EU), and how they merge or collide with official EU discourses of hygiene standards, on the one hand, and free trade, on the other, resulting in products that are good, clean, fair . . . and illegal.

I examine local food production in Latvia from Soviet times to the present, and demonstrate how the articulation of the political economy of food systems, bureaucratic controls on production, and moral ideals of consumption have resulted in a different positioning of foods and their procurement strategies as ethical in various historical and political periods. The Soviet planned economy resulted in scarcity at the level of raw materials that limited the availability of final products, yet also resulted in innovative ways of producing and obtaining them through underground channels. Knowledge and skills of food production, conservation, and preparation preserved in private kitchen spaces during the Soviet years are now being used to produce homemade products for sale within the EU. Yet while the current EU system offers an abundance of raw materials, it paradoxically also results in a scarcity of legally produced homemade products, due to strict quality standards and hygiene regulations in the processing phase that can limit innovation and access to markets. Despite the broader changes in the political economies of food production, producers and consumers in both periods have cultivated and created new social networks to obtain necessary goods and inputs that challenge official marketing channels and their relations to structures of power. I show how under both systems, informally or illegally obtained food products have borne the mark of authenticity, and bypassing the system has been equated with an ethical stance.

I will argue that in recent years in Latvia, a revival of interest in home-produced foods and culinary delicacies has been accompanied by an increasing awareness of the injustice of the required infrastructural investment and bureaucratic hurdles for small producers to comply with EU hygiene standards. As a result, homemade foods, valued as “good” and “clean” by consumers, continue to circulate through informal exchange networks, much as in Soviet times.[2] These are seen as “fair,” but are deemed illegal according to EU requirements. Positioning local informal networks as illegal, and thus unethical, obscures persistent inequalities regarding access to markets for the smallest home producers, and stigmatizes local practices and social networks as backwards without addressing the causes.

This article is based on ethnographic observations and in-depth interviews with artisan producers and consumers in Latvia in 2011 and 2012, and builds upon previous research conducted with organic producers and policy-makers from 2004 to 2012. I first describe a current revival of rural bread baking in Latvia as part of a local food network, and its relation to and differences from past Soviet networks. I then discuss the impact of current EU hygiene regulations and legal challenges for small-scale home producers to get their products to market, and analyze how this influences the persistence of local informal networks. Finally, I reflect on the consequences of criminalizing informal networks, rather than helping home producers legalize their products.

Good: Culinary Heritage Revivals

Baking bread at the Lejupes farm is a two-day process. It is something Valērija's family has done as long as she can remember.[3] Today her daughter-in-law Līga has taken over the baking, under her careful watch. On the first day, they revive the sourdough starter, mix some flour with hot water and salt, and stir it up in the *abra*, or wooden bread trough, that clearly shows its age. The soul of the bread is in the *abra*, they explain, and the *abra* is what gives each farm's starter its unique flavor. Both Valērija and her daughter-in-law shake their heads at the shame that some people have planted flowers in their bread troughs and no longer use them. "Not us. . ." they say.[4] At the end of each baking the remaining dough is scraped out of the *abra* and saved for use as a starter for the new batch. The *abra* never gets washed – any dough left after scraping it out hardens, and coats the wood like shellac – if it gets washed then the wood will mold, and the *abra* will be dead.

Once the mixture cools, they pound in the sourdough starter, using a long-handled wooden spoon. Then the mixture is left to ferment at least overnight, but longer in winter when it is cold. The next day, the slightly bubbly mixture is augmented with more flour, caraway seeds, sugar, and a tiny bit of live yeast, and kneaded by hand for half an hour in the same trough or "until your back is wet." The dough is left to rise another two and a half hours, if it is rising well. Then they light the fire in the oven, and let it heat for about an hour and a half. Once all the wood is burned through, they make the loaves and "shoot" them in with a long-handled bread peel; after 15 minutes they switch front loaves to back to ensure a more even baking and browning. At the end of the baking, they always load fresh wood in the oven to leave it with a "full belly."

This is the same basic recipe that was followed by Valērija's family in the Soviet years and by her daughter-in-law Līga now. Valērija had stopped baking for a few years in the early 1990s when she felt there was a decrease in flour quality and the bread "didn't come out right." When Līga wanted to start baking, they had to ask for sourdough starter from some friends who were still baking bread in the neighboring town, but she uses Valērija's recipe, the old *abra* and wood-fired oven. One consumer from Rīga told me, "That bread is so good, I can eat the whole loaf by myself – it just disappears in a day!"[5] Besides bread, they also sell milk, cream, and fresh cheeses, and occasionally other farm products or homemade delicacies, such as smoked meats and sausages. While Valērija and Līga are not part of any organized network such as Slow Food, they have received recognition in a recent wave of interest in reviving local culinary heritage practices in Latvia. The values embodied in the bread and its history are akin to those of culinary heritage movements throughout the world, recuperating family recipes, honoring "traditional"[6] baking practices, and experiencing taste and eating not only as a culinary pleasure, but also as a cultural phenomenon of social conviviality (Leitch 2003).

Fair: Historical Food Ethics

The Lejupes bread, however, also embodies a complex social history particular to post-socialist contexts, which imbues it with its "ethical" character, or the local sense of fairness in food consumption. While we were waiting for the bread to bake, Valērija told me her family's history of bread baking. Her family worked in the *kolhozs* (collective farm) in the Soviet years, but they kept baking bread at home. Otherwise, she told me, you would have to stand in line all night just to get half a loaf. When she was a child they lived in a different house, because in the first years after the war, private farms were collectivized, and they had had to leave. Their family was too poor to have a clock, but their neighbors had one. They would tell the neighbors when they put the bread in, and the neighbors would hang out a flag when it was time to take it out. Eventually they could return to their own house, which they had to share with other families, and were able to renovate the bread oven. Usually they baked rye bread that would last longer, but for special social occasions, like weddings, birthdays, and other celebrations, the traditional wheat sweet and sour-dough bread, *saldskābmaize*, described above, was a must.

Unlike in East Germany, where private bakers still existed throughout the socialist period, albeit with different rules (Buechler and Buechler 1999), in Latvia bread production was centralized, and home bakers did their baking secretly. The biggest problem was getting grain for flour. When I asked Valērija if her mother could get grain from the *kolhozs* for baking, Valērija laughed. "Get? She stole it! Stole the grain and took it to the mill!" Valērija exclaimed, with a wry smile and a note of pride in her voice. Her mother also knew the woman in charge of the granary, who sometimes would give her "elite seeds" if no grains were available. At first Līga and I were confused by this statement. "But where did you plant it?" asked Līga, knowing there was no room for grain production on small family plots. "No, she gave us the seeds . . . and we took them to the mill to grind them for flour!" Aaah, now we understood: rather than planting the grain that had been saved for seed, it was used as food. Favors, such as access to scarce products like grain, were known in the Soviet era by the term *blats*.[7] They became just as important ingredients in the bread as the flour itself, and using your connections to steal from the *kolhozs* was part of what maintained social relations.

In the Soviet Union, seed production was also centralized, and seeds were allocated accordingly to collective and state farms, in line with what Verdery (1991, 421) has described as the drive in socialism for bureaucrats to accumulate goods for the purpose of redistribution. The result was a tendency at all levels to hoard resources, thus creating an "economy of shortage" (Kornai 1980). Thus, in socialist economies, supply was scarce, rather than demand. Dunn (2008) has recounted numerous ways in which the Soviet Union controlled food production in order to make people more reliant on the state for their daily sustenance. The same of course was true of bread baking and other foods: private bakers constituted "a threat to the central monopoly on allocation" (Verdery 1991, 421), thus their access to grains was limited.

Yet rather than stopping production, these shortages promoted the informal economy and exchange relations (Verdery 1996). Valērija told me that due to the scarcity of grain, anyone who wanted bread for a special event would obtain all of the ingredients and bring them to her for baking. Through this, the demand for raw materials (and favors) was more evenly distributed, and not the sole responsibility of the baker. In order to fulfill the social need for bread on special occasions, grain saved as seed might be transformed back into grain. As Kloppenburg has said in his history of seed politics, "[S]eed is grain is seed is grain: the potential for either is there in each grain" (Kloppenburg 1988, 37). In the *kolhozs*, it was the keeper of the granary who decided the fate of each grain, but the hidden act of defiance remained a source of pride for the baker who obtained the highest quality illicit flour from the mill for her baking. This transformation of seed to grain allowed for the production of bread, the maintenance of social networks, and the creation of a moral and political stance against the state.

There is ample literature about the various functions that informal networks and the underground market had in socialist economies. Burawoy and Lukács (1992) and others have noted how workers on the shop floor devised plans of cooperation to hide each others' errors, and specific acts, such as pilfering, were considered acts of solidarity.[8] Consumption itself was also an ethical-political practice. Verdery has stated that under socialism "acquiring objects became a way of constituting your selfhood against a deeply unpopular regime" (Verdery 1996, 29). Similarly, Fehérváry (2002, 393–94) has noted that it was under state socialism that "mass consumer society and modern consumer subjectivities emerged." Furthermore, Fehérváry has noted that through critiquing state-produced products, "a more visceral relationship developed between consumption and political subjectivity during the socialist period in Eastern Europe than between consumption and citizenship in capitalist contexts (Auslander 1996; Cohen 2001)" (cited in Fehérváry 2009, 428).

In this context, specific products took on particular symbolism. Yurchak (2005) describes how Western music and jeans symbolized a Western "elsewhere" that need not have even been connected to the "real" West, but were significant markers of identity as they circulated through Soviet society. Exchange of food was a specific form of social currency in Russia, both during and after socialism, often used as barter or an insurance mechanism (Caldwell 2004; Ries 2009). Thus treading the thin line between social networks and rings of "corruption," these informal exchange networks were a necessary part of life under socialism. As Ledeneva (2009, 259) has observed, the system of favors known as *blat* in Russia was not considered illegal or criminal because it was seen as necessary for survival, "thus falling in the category of 'good' or 'ambiguous' corruption." Because the regime itself was seen as corrupt, she notes that the use of *blat* networks also became a form of resistance against the regime: "If *blat* corrupted the corrupt regime, can we refer to it as corruption?" (2009, 259). This logic positions these social networks of consumption under socialism as networks of resistance and solidarity, and illegal products, such as the homemade bread, as ethical.

Clean: European Economies of Purity

If in the Soviet Union restrictions in the supply chain came at the stage of raw materials, resulting in a scarcity of final products, in new EU member states like Latvia there is once again a scarcity of particular products as a result of strict hygiene regulations at the processing phase. Indeed, this issue of "cleanliness" as defined by hygiene standards for food processing was one of the most contentious points in EU accession in 2004 in Latvia and other new member states. Many producers of homemade breads, cheeses, jams, smoked meats, and other products feared that they would not be allowed to sell their homemade products. In response, the Ministry of Agriculture announced on the eve of EU accession in April 2004 that new "home producer" rules had been proposed:

> In Latvia the production of various food products in the home and their sale at markets is very popular. Who has not purchased a "*Jāņu siers*" (Midsummers' night cheese), smoked fish, homemade sausage, pickles, or honey? It is known that the main EU demand in relation to food is to ensure its *harmlessness*. This can be done with the use of the appropriate technologies and following hygiene regulations. The regulations [on home producers] have been developed in order to ensure the production and sale of these and other home-produced and highly valued products after entry into the EU. [emphasis added] (Latvian Ministry of Agriculture 2004)

Nevertheless, to the frustration of farmers, the home producer regulations were finally approved only two years later, in January 2006. Once adopted, the implementation of EU hygiene standards has created a new barrier for home producers to sell their products due both to the required paperwork and the associated expenses of implementation. An editorial in a Latvian news weekly in 2012 decried the fact that rural people are encouraged to become entrepreneurs, but that "the Latvian regulations for home producers reaffirm that we are 'holier than the pope' " (Benfelde 2012), and discourage this very possibility. In a 2009 study about home producers, even the Food and Veterinary Agency experts interviewed acknowledged that the legislation was still not exactly clear about who counts as a home producer versus a commercial producer, and that the standards for both groups do not actually differ too much (Latvian Rural Women's Association 2009). This means that home producers must meet nearly industrial standards, which translates into large infrastructural investments. For example, the regulations stipulate norms for the material of surfaces and walls, the type of lavatories, water supply, cooling facilities, transport, and a host of other technicalities required to ensure the cleanliness, or "harmlessness," of products. Implementing these standards is expensive in rural kitchens that may still not have running water, let alone tiled (washable) walls, separate entrances, and multiple indoor toilets. Furthermore, each required test for water quality, soil quality, or personal health costs money. A family that processes cheese on their goat farm complained to me in 2006 that, according to the regulations, they would have to build a separate bathroom for their grandmother, the cheese maker, to use during the hours she makes the cheese, even though the production space is in the same house where she lives (and has a bathroom).

It is not only home producers who were hit hard by the hygiene regulations. They also created problems for small processing facilities. The owner of a small organic bakery in 2006 said laughingly that he would have to write a book of memoirs just about the hygiene regulations. For example, the regulations require monitoring the flow of products, meaning that raw materials enter through one door, and final processed products must leave by another. The two should never cross. In assessing potential risks for food safety, one must be able to describe what to do in any calamity. The baker noted the absurdity of required detail with an example:

> Imagine, what do you do in the case that a loaf of bread falls on the floor – sometimes this happens – but what to do? You almost have to build a third door for this case, because you can't take it out the one where the raw materials are coming in, nor the other one where the finished product is going out . . . You can't dig a hole for it, because germs will get in there . . . so it is a big problem, what do you do?[9]

He also told me that the norms sometimes interfered with traditional baking processes. The bakers' association had to petition the authorities for the right to use wooden barrels for the fermenting of the sourdough and wooden surfaces for kneading the dough, rather than plastic or metal as required by the regulations: "They should try kneading dough on a cold metal surface. . .,"[10] he joked, bitterly.

Joking aside, for organic processing plants it is especially difficult to comply with all regulations, because they must be inspected both by the Food and Veterinary agency and the organic certifiers. If organic food is to be processed in a plant that is not dedicated only to organic products, organic processing must be separated in time and space from conventional food processing operations, meaning it takes place only during set hours of the day or week or in an entirely separate facility. Since there are still limited amounts of organic products to process, it is very costly for small companies to do this.

The bakery owner tried to do everything "by the books," but despite the fact that the bread was very popular among buyers, he came to the point of bankruptcy and had to sell his business a few years later. He is not alone in this fate. Within a few years, the only certified organic dairy and the only certified organic beef slaughterhouse and processing plant also went bankrupt. In 2012, even though a few slaughterhouses had certified organic slaughter times, according to an organic shopkeeper, there was still no certified facility to divide the carcass, thus it was still nearly impossible to get local certified organic beef into the market legally.

These bottlenecks also have the effect of displacing local sales. Many beef farmers now sell their live calves to intermediaries who arrive at the farm gate to take them for export to fattening facilities abroad, making selling food abroad more attractive than local sales. Despite the fact that many organic farmers in Latvia raise beef cattle, there is little or no local organic beef officially for sale in the country. The same is true for other organic producers, who increasingly see exporting raw materials within Europe as easier than negotiating local processing bottlenecks. Given these trends, many farmers and consumers in Latvia have expressed their fears that EU policies and standards are directly or indirectly aimed at eliminating competition for Western European companies (Kārkliņa, Maizītis, and Gulbe 2011).

Indeed the introduction of hygiene regulations is closely tied to the introduction of free trade. Standards, and their harmonization within free trade areas, are a means of global economic governance that often facilitate the entry of global capital, but create new barriers for small producers who lack infrastructure (Dunn 2003). Much has been written about the effect of EU standards on food practices and products in post-socialist contexts, noting how they reinforce inequalities among domestic producers and between small-scale local producers and incoming foreign investment. Mincyte (2011) has shown that Ministry of Agriculture documents in Lithuania explicitly acknowledged that hygiene standards were implicated in eliminating small-scale producers. Dunn (2003, 2008) has pointed out that joining EU hygiene standards in Poland and Codex Alimentarius in Georgia merely imposed norms that were nearly unattainable due to the lack of infrastructure, paving the way for foreign firms. Gille (2009) has decried the fact that European policy in old member states has differed dramatically from that aimed at new member states, which has promoted neoliberal reforms and structural adjustment similar to that pushed in developing countries in the 1980s.

These examples show that pressure in new member states for hygienic harmlessness can result in economic harm: hygiene standards are intimately linked with a push toward standardization and consolidation of industry promoted by neoliberal policies encouraging free trade. This link between hygiene standards and free trade, and the feared effects on small local producers, is crucial to understanding local conceptions of fair or ethical consumption, which I address in more detail below.

Illegal: Informal Social Networks

A regional newspaper wrote a story about the bread-baking tradition at Lejupes farm in 2011, but had to be cautious about what they wrote, because since joining the EU, the Lejupes bread was illegal once again. The old wood-fired bread oven was in a farm building that had not been renovated to meet EU hygiene standards, and due to this Līga had not registered as a home producer. They intended to do so, but the changes would require investments, so in the meantime they risked selling the bread illegally.

While customers need not procure the ingredients themselves any more, the Lejupes farm still sold its bread through informal social networks strikingly similar to those that existed under socialism. The loaves that Līga bakes now are much smaller than those Valērija used to bake, and are more expensive than bread sold in grocery stores. Because of this, locals still order the bread mostly for special occasions and when visitors come to town. But the farm also created new informal social networks, as word about their bread spread to tourists, who sometimes even go to the farm by the busload to buy products. And since the regional newspaper wrote about Lejupes, the entire newspaper staff orders the bread, specially delivered, still warm, to their city office every Friday. While Līga does the baking, it is her husband Jānis who takes orders and delivers the products. Each delivery or pickup involves at least a short chat and exchange of news; therefore, Jānis knows everything about everyone in town and is a social hub. Residents and tourists alike ask them where to find other service providers such as car mechanics or massage therapists, and even the journalists call him to get contacts with other town residents.

When I accompanied Jānis and his son on the delivery to the newspaper one Friday, the staff swarmed around the basket of bread and other farm goods, and products were snatched up within moments in an excited flurry. Newspaper staff commented that they buy from Jānis as much because of the bread as because of the connection with the farmer. “You’re buying not only the loaf of bread, but the farmers' attitude, their story . . . I'd rather support this concrete family, where I know they have a second small baby, than some mystical 'John' about whom I know nothing."[11]

All of the staff I interviewed said that certificates or health inspections did not matter to them, because this was a relationship based on trust. After all, they said, one did not have a certificate to cook in one's own home either. One customer, who said the bread tasted exactly like the one his grandmother used to bake, had been to the farm and said: "Imagine, if you went into the bake house, and it were all tiled – that would be repelling. There *should* be the old broom in the corner . . . if a mouse runs across the [room], that's fine, that's how it *should* be" [emphasis in original].[12] He emphasized that the bread was different each time – sometimes darker, sometimes softer, and this added to the appeal. Another consumer had considered the illegality of the bread, but had found a justification:

> the only minus is the taxes. . . . I don't know what they do [about income tax], but my husband and I discussed this at home. If Jānis goes and buys other products with the money – domestic items, then he pays the [sales] tax there . . . so I'm happier to support this specific family [than buy commercial bread].[13]

She emphasized that this would help the local economy. For most of these consumers, the illegality of the product was not an issue, but for those who did consider it, it was compensated by trust and personal connections. Supporting this farmer was considered fair in the local context.

The types of informal social networks Līga and Jānis have created and upon which they depend are not unique. Due to the long wait for the home producer regulations, many small farmers were resentful that they had to either stop producing items for the market or sell them under the table. In July 2004, only a few months after Latvia joined the EU, a journalist wrote that EU accession had facilitated the return of the gray economy, and that small business owners would "suffocate in their honesty" if they tried to pay all the taxes and other fees that favored large businesses (Paiders 2004). In fact, a 2011 study revealed that over 50% of Latvia's economy still falls into the gray or black zones (Točs 2011).[14] The 2009 study of home producers noted that over 81% of the surveyed people stated that selling value-added products was the only way to subsist in the countryside, yet only 28% of them had registered as home producers, citing bureaucratic hurdles as the main reason. Eighty-one percent also felt certain (55%) or suspected (26%) that those who had registered legally were worse off than those who had not (Latvian Rural Women's Association 2009). One home producer who had registered explained in a newspaper interview that she had since dropped the registration, because the extra expenses every year meant she couldn't break even (Misiņa 2012).

Consumers are sympathetic to the small rural producers' situation. Many would rather support local farmers, because their food is tastier, and it helps them make a living in the countryside. As a summer visitor in Līga and Jānis' rural town, I was dismayed to find that all the vegetables in the local store were imported from places like Spain or China, and all the bread and dairy products were commercially produced, despite the fact that every surrounding house produced some agricultural products. While those products may be cheaper, local consumers often preferred to buy from the farmer who set her vegetables out on the hood of her car across the street from the local store, or make an extra trip to Līga's to pick up milk, eggs, and other products. Līga observed that the economic crisis had not affected them at all. She said that people did not try to skimp on food – they might not buy shoes or clothes or other household items, but at craft markets where they sometimes sell their bread, people were increasingly buying more and different food products.

As in Soviet times, however, the Lejupes farm's main concern was that someone might denounce them to the authorities out of jealousy, prompting an inspection. Therefore, as their bread and other farm products became more popular, in 2012 they began to rebuild their bake house in accordance with EU standards so that they may sell their bread legally. They were reluctant to take out a loan, however, having seen that many official and renowned bakeries went bankrupt due to competition with baked-in-the-store supermarket bread, and farmers who took out loans before the crisis were struggling to repay them or losing their businesses. They also hoped to be able to negotiate with the authorities about the style of the building, as they preferred to preserve a more rustic look than a sterile industrial one. "When I walk in here, I have to feel the history of the place, otherwise I can't bake," said Līga.[15]

We Have Never Been Legal

The continuity of informal networks should come as no surprise, as it is directly related to the types of political control exercised in the economy. Dunn (2003) predicted already before EU accession that informal networks would continue to persist in the Polish meatpacking industry in response to the high costs of standardization that were spurring consolidation and foreign takeover of the industry. Similarly, Metzo (2001, 555) has argued that *blat* networks continued in the post-socialist period in Russia largely due to the uncertainty of economic reforms, and morphed into a "local market economic structure" in the absence of formalized markets in remote areas. Furthermore, Stan (2012, 65) argues that from a political economy perspective, all informal exchanges are embedded in power relations, and therefore they are "neither legacies of the past nor prefigurations of the future, but rather a direct consequence of current political and economic transformations."

In contrast, Ledeneva (2009) has argued that once the economy of shortage in Russia was replaced with readily available markets of goods, *blat* lost its importance for everyday consumption, and is now used more for attaining special favors from people in power. While this may have been true in Latvia in the 1990s, and may continue to be true in Russia or other post-socialist countries that did not join the EU, we see that in Latvia EU hygiene regulations resulted once again in a scarcity of legally produced homemade products, and thus re-instituted the necessity for informal exchange networks in the eyes of producers and consumers. One woman in Latvia noted the irony in the continuity of these informal networks from the Soviet Union to the European Union: "I sold smoked pork under the table then, because the [extra] pig was raised at home; now I have to sell smoked pork under the table again, because it's been smoked at home."[16]

I would like to suggest that rather than merely lingering out of habit or due to lack of infrastructure, the current circulation of illicit food items in new member states reflects a pushback against the EU bureaucratic regime, much like the circulation of Western goods such as jeans and rock music in the Soviet era (Yurchak 2005). In Lithuania, Mincyte (2009) has argued that informal networks for the sale of raw milk are more than simply resistance to EU standards, but operate as an active means for female producers and consumers to take control of their bodies and their ways of life. While most such acts of defiance remain hidden from the public eye, there has recently also been public protest against agriculture regulations. In January 2012, a farm in Latvia was accused of the illegal sale of tomato seeds not in the European Common Catalogue.[17] As a result, the farm experienced record sales and unprecedented solidarity from consumers, who left their phone numbers in case the farm needed any help. The farmers were accompanied to their hearing by a group of vociferous demonstrators and members of their gardening club who helped them defend their rights (Aistara 2014b). While several editorials had appeared about the discriminatory effect of home producers' laws (Misiņa 2012), in 2012 there was not yet an active lobby group or NGO defending their rights. Rather, people "vote with their wallet," continuing to buy from valued neighbors, regardless of whether they are legally registered or not. The Food and Veterinary Department warned that if illegal sellers were caught, however, they may face fines of up to 500 Lats (700 Euro) (Misiņa 2012).

There is a further consequence, however, besides the mere illegality of the homemade products and those producing or consuming them. Purchasing goods through informal networks is also interpreted by some as a sign of backwardness and an impediment to becoming modern (Ledeneva 2009). In a recent editorial in Latvia encouraging businesses to purchase local goods rather than global ones for employee holiday gifts as a part of corporate social responsibility, the author lamented that the two biggest "crimes" committed by Latvian producers are not paying taxes and accepting or giving bribes or favors (Minskere 2012), indicating that such informal practices disqualify them from the type of ethical consumption the author would like to promote. Similarly, an NGO worker promoting responsible consumerism told me the old *blat* networks are a leftover from the Soviet era that should be eliminated now that Latvia has joined the EU.

Such discourse of comparison to Western norms mirrors a long history of neo-orientalist positioning of eastern Europe as the backwards Other of the modern western Europe (Buchowski 2006). As Klumbyte (2009) has observed in recent Lithuanian debates over the comeback of "Soviet" sausage, those who wished to consume the "Soviet" sausage were assumed to want to retreat backwards to a communist past. They were called "turnips" and assumed to represent an obstacle to the country's path forward to Europe. Yet Klumbyte (2010) shows that rather than reflecting backwardness, nostalgia for the Soviet also expresses a powerful critique of the present. This division of consumers and producers into modern and backwards parallels Latour's description of the presumed rupture between the past and modernity: "[T]ime's arrow is unambiguous: one can go forward, but then one must break with the past; one can choose to go backward, but then one has to break with the modernizing avant-gardes, which have broken radically with their own past" (1993, 69). In a similar manner, continuing to support informal networks poses a trap for the "turnips," as this presumes them to be against the modernity represented by the EU.

The criminalization of small producers unable to attain hygiene standards, and the demarcation of both the producers and consumers who buy from them as backwards or pre-modern, does nothing to address the underlying structural causes of competing economic pressures in the countryside that prevent people from entering legal or "modern" systems. In fact, it may do more to perpetuate the system, as this criminalizing discourse may make consumers and producers retrench to their customary informal market relations for obtaining local foods, as these represent a stance against the inequalities or injustices that have produced the problem.

Despite the stated aim of food standards to "level the playing field," as Dunn (2003, 1507) has elaborated, "the cost of regulation is not the same for all producers." This is particularly true for small-scale home producers of everyday products such as jam, baked goods, dairy products, and others: the costs for meeting infrastructure-related standards will be hardest to bear for those who most need to sell these products in order to subsist in the countryside. Social inequality in Latvia and Lithuania are the highest in the EU, and long-term unemployment is a persistent problem in rural areas (Masso et al. 2012). The geographical area with the fewest registered home producers coincides with the poorest region of the country (Latvian Rural Women's Association 2009). To criminalize these small producers who cannot meet the standards, rather than to devise alternative policy mechanisms to facilitate their transition to "legality," is to deny the economic inequalities and hardships out of which they are borne, and effectively prevent them from ever becoming legal.

For Latour (1993), modernity, or the perfect separation of nature from culture, is unattainable because it depends on the denial of the very nature-culture hybrids that it continues to produce, resulting in a constant effort of "purification" to eliminate them. Similarly, the criminalization of home producers will not make them disappear, but rather create conditions under which they are more likely to appear. The home producers who have not been legal under either system may thus be doomed to also never be legal. Following Latour, as they have never been legal, they have also never been modern (and never will be) (Latour 1993).

Conclusion: History, Harmlessness, and Hygiene

If the Soviet Union perpetuated an economy of scarcity, the European Union maintains an economy of purity. While in the Soviet years a lack of raw materials restricted production, in the EU, hygiene regulations restrict production. Both result in a scarcity of legal homemade food. Thus producers and consumers also revert to their accustomed forms of social organization through relying on informal networks that create an imagined community of resistance against unfair regulations. In Soviet times, people were willing to pilfer ingredients for baking bread for special occasions, as much for the stance it demonstrated against state control as for the bread itself. Today, people are willing to pay more for local, illegal, bread because of the social ties it creates to a family maintaining a tradition that is not valued by the current system. The consumption of illegal goods thus serves as an ethical and moral stance, and solidarity is built through networks of dissent. Individuals continue to choose buying local homemade goods through their informal networks: it may be illegal, but it is good, clean, and fair.

This response is also ironic, however, because it reinforces the reliance on individual consumption to bring about social change, as embodied in principles of neoliberal free trade, rather than addressing the underlying structural causes of inequalities in access to markets. Regulations must be adapted in ways that permit the most marginalized home producers to achieve "harmlessness" and enter legal markets, through which they may both contribute to and benefit from state services, rather than criminalizing and stigmatizing producers, consumers, and their social networks.

These complex histories remind us that "food ethics" are not something absolute, constant, and unchanging, but rather relative, situational, and fundamentally political. The illegal becomes ethical when what is valued becomes contraband, and food practices become mechanisms through which to mediate relations to the state as well as to build networks of solidarity. Rather than diminishing in the economic crisis, these networks may prove to be the survival skills necessary to face it, much as in previous historical eras.

Acknowledgments

A previous version of this paper was presented at the European Association for Social Anthropology meetings in Nanterre in 2012, where it benefitted from comments from organizer Cristina Grasseni and discussant Heather Paxson. I am grateful to Diana Mincyte, Ulrike Plath, Hadley Renkin, Karen Hébert, Laura Sayre, and two anonymous reviewers for comments on earlier drafts.

Disclosure Statement

No potential conflict of interest was reported by the author.

Notes

1. While I use Slow Food's slogan "Good, Clean, Fair" to structure the paper, I do not intend this as a commentary on the Slow Food movement.
2. This article reflects the situation at the time of writing in October 2012. A few positive changes had taken place by the time of going to press in June 2015, including a new hybrid form of "direct buying circles" initiated by consumers who wish to uphold direct relationships with producers yet also try to find ways to legalize these systems. Līga and Jānis have also since legalized their bakery. Unfortunately, I am not able to cover these changes in depth in this article.
3. All names have been changed and all translations are my own. This and all statements by Valērija and Līga in this section: interview, Lejupes farm, 17 August 2011.
4. Their disappointment may be even greater to learn that these same antique wooden bread troughs are now sold on the Internet as "found vintage dough bowls" by the Pottery Barn chain store in the United States, complete with a video instructing users how to arrange seasonal produce, eucalyptus sprigs, and votive candles.
5. Consumer interview, Rīga, 25 October 2011.

6. My use of the term "traditional" here does not imply authenticity in the sense of unchanging, age-old practices, as these recipes have changed over time, and the Soviet version itself could be contested by some as "invented." See Aistara (2014a) for more on the construction of authenticity in renewed culinary traditions in Latvia.
7. This is the same term as *blat* in Russian, cited in articles below.
8. As pointed out by an anonymous reviewer, pilfering was not always seen as a just practice by everyone in Soviet times, but was rather the subject of great negotiation.
9. Interview with baker, organic bakery, 19 October 2005.
10. Interview with baker, organic bakery, 19 October 2005.
11. Interview with anonymous consumer, city newspaper office, 2 November 2012.
12. Interview with anonymous consumer, city newspaper office, 2 November 2012.
13. Interview with anonymous consumer, city newspaper office, 2 November 2012.
14. The gray economy includes transactions where registered businesses do not follow all regulations, whereas the black economy includes non-registered businesses.
15. Interview with Līga, 4 November 2012.
16. Under Soviet regulations, people were allowed to keep one pig in their home plots, but tried to get around that. Interview in Rīga, 2 November 2012.
17. Similar to hygiene regulations, EU seed laws have a limiting effect on the types of seeds that can be sold by small producers on the market.

References

Aistara, G. 2014a. "Authentic Anachronisms." *Gastronomica: The Journal of Food and Culture* 14 (4): 7–16. doi:10.1525/gfc.2014.14.4.7.

Aistara, G. 2014b. "Latvia's Tomato Rebellion: Nested Environmental Justice and Returning Eco-sociality in the Post-socialist EU Countryside." *Journal of Baltic Studies* 45(1): 105–130. doi:10.1080/01629778.2013.836831.

Auslander, L. 1996. *Taste and Power: Furnishing Modern France*. Berkeley: University of California Press.

Benfelde, S. 2012. "Īpašie Noteikumi Mazajam Cilvēkam." *Ir*, July 31.

Buchowski, M. 2006. "The Specter of Orientalism in Europe: From Exotic Other to Stigmatized Brother." *Anthropological Quarterly* 79 (3): 463–482. doi:10.1353/anq.2006.0032.

Buechler, H., and J.-M. Buechler 1999. "The Bakers of Bemburg and the Logics of Communism and Capitalism." *American Ethnologist* 26 (4): 799–821. doi:10.1525/ae.1999.26.4.799.

Burawoy, M., and J. Lukács. 1992. *The Radiant Past: Ideology and Reality in Hungary's Road to Capitalism*. Chicago, IL: University of Chicago Press.

Caldwell, M. L. 2004. *Not by Bread Alone: Social Support in the New Russia*. Berkeley: University of California Press.

Carrier, J., and P. Luetchford, eds. 2012. *Ethical Consumption: Social Value and Economic Practice*. Oxford, MS: Berghahn Books.

Cohen, L. 2001. "Citizens and Consumers in the U.S. in the Century of Mass Consumption." In *The Politics of Consumption: Material Culture and Citizenship in Europe and America*, edited by M. Daunton and M. Hilton, p. 203–222. Oxford: Berg Press.

Cook, I. 2000. "Cultural Geographies in Practice: Social Sculpture and Connective Aesthetics: Shelley Sacks's 'Exchange Values'." *Cultural Geographies in Practice* 7 (3): 337–343. doi:10.1177/096746080000700306.

Dunn, E. 2008. "Postsocialist Spores: Disease, Bodies, and the State in the Republic of Georgia." *American Ethnologist* 35(2): 243–258. doi:10.1111/j.1548-1425.2008.00032.x.

Dunn, E. C. 2003. "Trojan Pig: Paradoxes of Food Safety Regulation." *Environment and Planning A* 35(8): 1493–1511. doi:10.1068/a35169.

Fehérváry, K. 2002. "American Kitchens, Luxury Bathrooms, and the Search for a 'Normal' Life in Postsocialist Hungary." *Ethnos* 67(3): 369–400. doi:10.1080/0014184022000031211.

Fehérváry, K. 2009. "Goods and States: The Political Logic of State-Socialist Material Culture." *Comparative Studies in Society and History* 51 (2): 426–459. doi:10.1017/S0010417509000188.

Gille, Z. 2009. "Globalizing Paprika: Food Governmentalities in the Postsocialist European Union." In *Food and Everyday Life in the Postsocialist World*, edited by M. Caldwell, 57–77. Bloomington: Indiana University Press.

Guthman, J. 2004. *Agrarian Dreams: The Paradox of Organic Farming in California*. Berkeley: University of California Press.

Kārkliņa, G., R. Maizītis, and A. Gulbe. 2011. "Latvijas Dīvainais Pārtikas Tirgus. Neizdarība Vai Eiropas Diktāts?" Accessed February 20, 2013. http://www.kasjauns.lv/lv/zinas/56905/latvijas-divainais-partikas-tirgus-neizdariba-vai-eiropas-diktats

Kloppenburg, J. 1988. *First the Seed: The Political Economy of Plant Biotechnology, 1492-2000*. Cambridge: Cambridge University Press.

Klumbyte, N. 2009. "The Geopolitics of Taste: The 'Euro' and 'Soviet' Sausage Industries in Lithuania." In *Food and Everyday Life in the Postsocialist World*, edited by M. Caldwell, 130–153. Bloomington: Indiana University Press.

Klumbytė, N. 2010. "The Soviet Sausage Renaissance." *American Anthropologist*, 112 (1): 22–37. doi:10.1111/j.1548-1433.2009.01194.x.

Kornai, J. 1980. *Economics of Shortage*. Amsterdam: North Holland Press.

Latour, B. 1993. *We Have Never Been Modern*. Cambridge, MA: Harvard University Press.

Latvian Ministry of Agriculture. 2004. "Mājās Ražotus Produktus Varēs Tirgot, Ievērojot Noteiktas Prasības." *Press release*, Accessed January 19, 2014. http://www.agropols.lv/zinasprint.php?id=61080

Latvian Rural Women's Association. 2009. Lauku Mājražotāju Uzņēmējdarbību Kavējošo un Veicinošo Faktoru Izpēte Un Veicinošo Aktivitāšu Apzināšana un Ierosināšana. Rīga: Latvijas Lauku Forums.

Ledeneva, A. 2009. "From Russia with Blat: Can Informal Networks Help Modernize Russia?" *Social Research* 76 (1): 257–288.

Leitch, A. 2003. "Slow Food and the Politics of Pork Fat: Italian Food and European Identity," *Ethnos* 68 (4): 437–462. 10.1080/0014184032000160514.

Masso, J., K. Espenberg, A. Masso, I. Mierina, and K. Philips. 2012. "Growing Inequalities and Its Impacts in the Baltics." In *GINI Growing Inequalities Impacts*. Amsterdam: Amsterdam Institute for Advanced Labour Studies.

Metzo, K. 2001. "Adapting Capitalism: Household Plots, Forest Resources, and Moonlighting in Post-Soviet Siberia." *GeoJournal* 54: 549–556. doi:10.1023/A:1021792929051.

Mincyte, D. 2009. "Self-Made Women: Informal Dairy Markets in Europeanizing Lithuania." In *Food and Everyday Life in the Post-Socialist World*, edited by M. Caldwell, 78–100. Bloomington: Indiana University Press.

Mincyte, D. 2011. "Subsistence and Sustainability in Post-Industrial Europe: The Politics of Small-Scale Farming in Europeanising Lithuania." *Sociologia Ruralis* 51(2): 101–118. doi:10.1111/soru.2011.51.issue-2.

Minskere, L. 2012. "Atbalstīt Lokālo, Nevis Globālo." *Ir*, October 8.

Misiņa, I. 2012. "Mājražotāji Neņem Miljonu." *Latvijas Avīze*, March 12.

Moberg, M. ed. 2010. *Fair Trade and Social Justice: Global Ethnographies*. New York: New York University Press.

Paiders, J. 2004. "Legālais Bizness Kļūst Par Pelēko." *Neatkarīgā Rīta Avīze*, July 5.

Ries, N. 2009. 'Potato Ontology: Surviving Postsocialism in Russia.' *Cultural Anthropology* 24 (2): 181–212. doi:10.1111/cuan.2009.24.issue-2.

Stan, S. 2012. "Neither Commodities nor Gifts: Post-Socialist Informal Exchanges in the Romanian Healthcare System." *Journal of the Royal Anthropological Institute* 18: 65–82. doi:10.1111/jrai.2012.18.issue-1.

Točs, S. 2011. "Melnā un Pelēkā Ekonomika Virs 50%." *Diena*, May 16.

Verdery, K. 1991. "Theorizing Socialism: A Prologue to the 'Transition'." *American Ethnologist* 18(3): 419–439. doi:10.1525/ae.1991.18.3.02a00010.

Verdery, K. 1996. *What Was Socialism? What Comes Next?* Princeton, NJ: Princeton University Press.

Yurchak, A. 2005. *Everything Was Forever, Until It Was No More: The Last Soviet Generation*. Princeton, NJ: Princeton University Press.

GEOGRAPHIES OF RECONNECTION AT THE MARKETPLACE

Renata Blumberg

Since 2009, Vilnius' urban landscape has been transformed by the rapid growth of farmers' markets, mirroring tendencies in other parts of Europe and Northern America. Existing research has found that farmers' markets foster *social* and *spatial embeddedness*, meaning locally based relationships characterized by trust and reconnection. In contrast to these findings, I argue that social and spatial embeddedness are not guaranteed outcomes of market transactions in Vilnius farmers' markets. To explain this discrepancy, I argue that farmers' markets should be understood as unbounded places that are relationally constructed with other retail places, and produced by historical trajectories of production and consumption.

In 2009, the urban landscapes of Lithuania witnessed a dramatic increase in the number of farmers' markets. In a little over a year, the number of farmers' markets in two of Lithuania's largest cities, Kaunas and Vilnius, grew from just a few weekly markets to over 40, functioning for the most part on a yearlong basis. This trend is consistent with the growing popularity of farmers' markets in Europe and Northern America in the recent past. Farmers' markets have become emblematic of new efforts to foster alternatives to industrially produced, processed, and marketed food; they are crucial meeting points for "alternative food networks," direct-to-consumer supply chains that circumvent conventional modes of distribution and often operate under organizational principles that prioritize reconnecting consumers with producers (Goodman, DuPuis, and Goodman 2012). In Europe and Northern America, policy-makers, activists, and scholars have confirmed that the impressive growth of farmers' markets is explained in part by their success in satisfying a need for reconnection between producers and consumers (Karner 2010; Kneafsey et al. 2008).

Scholars have found that the face-to-face interactions between consumers and vendors at farmers' markets foster *social* and *spatial embeddedness*, meaning locally based relationships characterized by trust and reconnection (Feagan and Morris 2009). While the concept of embeddedness has been variously defined and utilized in academic work (Peck 2005), I use the concepts of social and spatial embeddedness as they have been developed in the literature in agro-food studies (Feagan and Morris 2009) to examine the relationships between producers and consumers in Vilnius' farmers' markets. I demonstrate that in contrast to existing findings on farmers' markets, social and spatial embeddedness are not guaranteed outcomes of market transactions in my case study sites. To explain the discrepancy between my findings and existing research, I argue that farmers' markets should be understood as place-making projects, which are relationally constructed with other retail places, and produced by particular historical trajectories of production and consumption. More specifically, thinking relationally about farmers' markets requires understanding how they are connected with existing retail places, which in Vilnius includes not only supermarkets but more importantly the longstanding and popular public markets where a certain kind of market culture has developed over time.[1]

In this article, I employ a methodological framework based upon a relational comparison (Hart 2002) of public and farmers' markets. Specifically, I utilize an empirically informed theoretical perspective, which is characterized by an open and unbounded understanding of place (Massey 2005). In order to assess evidence of social and spatial embeddedness, I analyze material from public and farmers' markets; informal and formal interviews with farmers, consumers, and farmers' market organizers; and promotional texts and media (Blumberg 2014).[2] These materials were coded and then analyzed in an iterative process. The research I undertook on markets was part of a larger project on alternative food networks in Lithuania that took place over the course of 11 months starting in late 2009 and ending in 2013.

This article is organized into five sections. In the first section, I review the research on farmers' markets and locate my research in this literature. In the second section, I develop a theoretical approach designed to address gaps in our understanding of farmers' markets. In the third and fourth sections, I focus on public markets and farmers' markets, respectively, first providing a context for their development and then analyzing and comparing the interactions that occur in both.[3] I conclude with reflections about the implications of my research for policy formation and scholarly research.

Making Space for Farmers' Markets

In Northern America and parts of Western Europe, farmers' markets have experienced a revival in popularity in recent years among the general public, and they have also become central to scholarly inquiry and public policy on consumption and rural development (Brown 2002; Holloway and Kneafsey 2000; Sage 2003). Since World War II, farmers' markets, as well as public markets more generally, experienced a period of decline that was in part spurred by the explosive growth of supermarkets. The near obliteration of farmers' markets for several decades has contributed to the novelty and excitement surrounding the establishment and growth of new farmers' markets during the last decades of the twentieth century. In the United Kingdom, for example, farmers' markets "represent a new and distinctive dimension to the somewhat 'placeless' foodscape of contemporary Britain" (Holloway and Kneafsey 2000, 286). More than just a fleeting novelty, scholars contend that "FM [farmers' markets] and other new retail forms are bringing together ever-larger numbers of producers and consumers within a fundamentally different type of relationship than that found in conventional supply chains" (Sage 2007, 150).

While research has demonstrated that the creation of farmers' markets has potential positive impacts on the environment, health, and local economic development (Brown and Miller 2008; Holben 2010), scholars have also sought to understand why consumers and producers are increasingly shopping or selling at farmers' markets. It is often argued that the turn away from conventional modes of food distribution is part of a broader movement to recover the moral dimensions of economic activity (Sage 2003). In contrast to conventional supply chains that work by fostering distance and ignorance between spheres and production and consumption, farmers' markets help bring together producers and consumers in direct, face-to-face relationships. In his analysis of farmers' markets in Ireland, Kirwan explains that "the producers and consumers concerned are engaging in face-to-face interaction in order to create conventions of exchange which incorporate spatial and social relationships that can replace 'uniform standards' with individualised judgement, thereby helping to overcome uncertainty" (2006, 303). Scholars have categorized this re-incorporation of spatial and social relationships into the economic sphere as forms of *spatial* and *social embeddedness* (Feagan and Morris 2009).

While social embeddedness denotes relations of trust, reconnection, and responsibility between consumers and producers (Sage 2003), spatial embeddedness locates these relationships spatially within the local scale (Feagan and Morris 2009). The face-to-face interactions at farmers' markets help forge social embeddedness by allowing producers to respond to individual consumer questions and needs (Kneafsey et al. 2008), and by helping producers build the trust necessary to ensure a low-risk market outlet devoid of the official contractual obligations that characterize conventional supply chains. By providing an outlet for local food, farmers' markets not only ensure spatial embeddedness but they are also considered to be key parts of the socio-economic infrastructure needed to rebuild local food systems (Gillespie et al. 2007).

Farmers' markets and the local food systems they support are not universally praised. While farmers' markets may be defined by their spatial embeddedness and the "local" scale can be the harbinger of progressive ideals, critics argue that the "local" may be constituted by reactionary politics as well (DuPuis and Goodman 2005). Moreover, in the Anglophone world, the "local" scale does not have a rigid boundary, but farmers' markets might create such boundaries to ensure spatial embeddedness. One effect of this practice is that rigid and arbitrary definitions of the "local," such as a set circumference around a city, may exclude marginalized farmers who live farther from wealthy urban centers. Farmers' markets have also been critiqued for fostering an exclusive form of social embeddedness by catering to wealthy and highly educated consumers (Slocum 2008). In the United States, Rachel Slocum has found that "as vehicles to augment grower incomes through better prices, some markets cater implicitly (organic-only, location, music, classes) to a well-off, educated and often white demographic" (Slocum 2008, 851). Further research has revealed the limitations inherent in any market-based solution to the problems of the food system (Alkon and Mares 2012). Although these critiques have cast a shadow on laudatory claims about farmers' markets, few scholars have investigated how the historical sociospatiality of farmers' markets as places influences the relationships they sustain. For example, the claims that farmers' markets are providing a venue for the cultivation of social and spatial embeddedness assume that existing retail places did not offer such forms of embeddedness. While this may be the case in some places, such as the United States, post-Soviet space offers another sociospatial context, with its own history of marketplaces, which plays a role in shaping even *new* farmers' markets.

In the following section, I draw upon the work of Doreen Massey and others to help account for the importance that history and space have in the making of farmers' markets as places. This has ramifications for their capacity to foster social and spatial embeddedness.

Rethinking Markets as Places in Lithuania

While commonplace understandings of place characterize places as fixed entities with defined borders and stable characteristics, the approach I utilize to rethink farmers' markets as place-making projects builds upon an understanding of places as heterogeneous multiplicities that are open and unbounded and always in the process of being made (Dzenovska 2013; Massey 2005). What makes a farmers' market a place are the multiple practices, meanings, and material and immaterial elements that come together and forge a dynamic, disparate configuration. An open and unbounded understanding of place does not presuppose complete spatial fluidity; rather, place making is a structured and structuring process that articulates with local histories and relations across space (Hart 2002). Indeed, particular historical trajectories are at work in the making of farmers' markets in contemporary Lithuania. Trajectories of production and consumption have been radically transformed since the break-up of the Soviet Union and the implementation of "transition" policies in the 1990s. Subsistence farming and allotment gardening were widespread practices in the Soviet Union, but the reforms of the 1990s, including privatization and decollectivization, led to the creation of an agricultural sector dominated by small-scale farms (Alanen 2004). Although the new small-scale farmers were able to sustain themselves in the difficult conditions of the 1990s by drawing upon an existing knowledge basis and forming alternative food networks, many experienced marginalization with European Union (EU) accession (Aistara 2014; Knudsen 2013; Mincyte 2011a). For consumers, practices have evolved from a brief period of fascination with the newly emerging Western-style supermarkets to include preferences for buying food directly from farmers and even nostalgia for Soviet-era products.[4] These two intertwined trajectories of production and consumption help make farmers' markets heterogeneous places in Lithuania.

Equally important in the making of farmers' markets as places are relations across space, such as between the rural and the urban, and between farmers' markets and other retail places. While the imaginary of the "local" (*vietinis*) in Lithuanian encompasses spatially bounded relations between rural and urban places, in contrast to the United States and the United Kingdom, the boundaries of the local are not ambiguous or contentious; the local has come to be equated with the territory of Lithuania. The connection between the local and the territory of the nation-state, rather than an arbitrary circumference around a city or a sub-national region, carries with it certain implications that have to do with national belonging and its contentious construction within a Europeanizing and globalizing sociospatial context (DeSoucey 2010; Mincyte 2011b).

Until the arrival of farmers' markets, the urban, public markets were important meeting points for the trajectories of consumption and production that brought together the rural and urban, forging places where social and spatial embeddedness could be practiced. Although Vilnius' public markets have existed and survived throughout multiple different political regimes, they blossomed during the 1990s, not only with farmers selling their own produce but also with resellers trying to make a livelihood in difficult economic circumstances. These resellers are also part of the trajectories that make public markets as places. While public markets do not claim to be farmers' markets, they play a significant role in the place making that occurs at farmers' markets because places are produced through their relations to other places, such as other places of retail. In most of the United States and parts of Western Europe, where public markets have only had a marginal presence in the past few decades, farmers' markets offer an interpersonal shopping experience that is attractive to those consumers who seek an alternative to comparatively impersonal supermarkets (Sage 2007). As a result, farmers' markets have become new and hopeful places for disillusioned consumers. For the farmers who cannot meet the standards required for selling in supermarkets, farmers' markets help ensure livelihoods. In Vilnius, by contrast, farmers' markets have emerged in relation to supermarkets *and* existing public markets. Therefore, not only are interpersonal connections between farmers and consumers nothing new, they also carry with them certain meanings and practices that have become very much a part of farmers' markets. The purpose of the following section is to delineate the kind of market culture that has emerged in the public markets of post-socialist Lithuania. I analyze two of Vilnius' prominent public markets by focusing on how recent historical trajectories have come together to make these marketplaces and by examining the kinds of interactions between vendors and consumers that take place there. This analysis forms a framework for understanding the development of farmers' markets in Vilnius, Lithuania.

Trajectories Making Places: "Turgus"

With the collapse of the Soviet Union and the easing of border controls, open-air and public markets (in this section, simply "market" or "*turgus*") filled a retail vacuum until new supermarkets and shopping malls began to dot Vilnius' urban landscape. Although their prominence has faded since the 1990s, markets are still important places for consumers to purchase food and for farmers to sell directly to consumers. In fact, a recent study revealed that 77% of Lithuanian residents shop at markets for food (Mikelionytė, Lukošiutė, and Petrauskaitė 2010). Despite their popularity, markets have not been universally perceived as respectable places. During the Soviet era, they were crucial places for the functioning of second economies (Sik and Wallace 1999), but officially "they were considered either as remnants of an outdated and unnecessary form of commerce or as a dangerous challenge to the socialized retail sector, as places where profit-making was combined with criminal activity such as speculation, pick-pocketing, or the reselling of smuggled and stolen property" (Sik and Wallace 1999, 697). As state control over trading weakened in the 1980s, public markets began to thrive, and following the collapse of the Soviet Union, they grew throughout post-Soviet space (Aidis 2003; Hohnen 2003; Hüwelmeier 2013; Mažeikis 2004; Polese and Prigarin 2013; Yükseker 2007). They became places to procure everyday consumer products, in addition to being places for the sale of counterfeit and illicit goods. While their chaotic and unruly characteristics and their connection with the Soviet past was a cause for frustration among public officials who believed such markets should not exist in a "civilised market economy" (Aidis 2003, 469), these officials also believed that the establishment of a formal retail sector would cause markets to become obsolete.

Despite efforts to eradicate or control them, however, markets did not disappear. Instead, the orderly chaos of the markets led to the growth of a market culture resembling the bazaar cultures widely studied by anthropologists (Geertz 1978). In bazaars, the absence of standards, fixed prices, or other official methods to ensure quality motivates consumers to prioritize establishing relationships with vendors, who might then offer special deals, preferential access, or simply highly coveted information. This situation prompts buyers and sellers in the bazaar economy to become "intimate antagonists" (Geertz 1978, 32), brought together by mutual necessity and interest, but in a situation in which both parties are motivated to get the best deal. Successful interactions breed "clientelization" (Geertz 1978, 30), the on-going patronage of the same sellers by consumers who favor reliability over trying something new and more risky. In the bazaar economy, qualities of social embeddedness like trust or responsibility are not guaranteed through face-to-face interactions, but must instead be cultivated. Similarly, spatial embeddedness must also be cultivated through social relations with vendors who are trusted to sell locally produced goods, either their own or those procured from others.

The resemblance that markets of the post-Soviet sphere may bear with bazaars worldwide notwithstanding, scholars of the region have argued that post-Soviet markets have had a path-dependent development trajectory (Czakó and Sik 1999). Whereas markets took advantage of a niche produced through deficiencies in the centrally planned economy, after 1991, they persisted and even grew with the opening of borders, the undeveloped formal retail sector, and simultaneous consumer demand. Moreover, economic transition policies in the 1990s produced high unemployment, but working people also struggled with low and insufficient salaries. As a result, post-Soviet societies hosted a ready supply of potential traders, who sought to maintain their livelihoods in an increasingly unstable environment. For farmers without reliable processing or other retail outlets, markets were important trading places, and the practice of selling at markets was not unfamiliar. Although trading may have been seen as immoral in the Soviet period, as Czakó and Sik (1999) explain, small-scale trading in Central and Eastern Europe has a strong cultural legacy. This legacy, as well as the human capital that came along with it, decreased the transaction costs for starting and maintaining markets or for becoming a trader (Hohnen 2003).

In summary, the importance of markets in the 1990s was made possible by the coming together of various historical trajectories. For consumers, markets provided a place to shop for highly valued but scarce consumer goods in the Soviet period, and in the post-Soviet period, they grew in significance as the state-supported retail sector crumbled. For all sorts of vendors, whether they sold used automobiles or vegetables from their kitchen gardens, markets provided a meaningful source of livelihood. Yet, markets continued to be contradictory places, both valued and regarded with suspicion. For consumers, the twofold characteristic of the markets as places where one can potentially bargain for the best deal *and* places where one can be cheated are common perspectives. Despite efforts to control, contain, or eliminate markets, and despite the contemporary dominance of a formal retail sector, markets are still important places of consumption, especially for food products (Mikelionytė, Lukošiutė, and Petrauskaitė 2010).

Vilnius Markets

Within the city of Vilnius, the two biggest markets are Halės Turgus and Kalvarijų Turgus, located on opposite sides of the center of the city. While both markets offer similar consumer goods, such as processed and fresh food and clothing, product variety is much greater in Kalvarijų Turgus, which fills a whole city block and also has a section devoted to used goods. Both markets include indoor market spaces, outdoor stalls, and entryways that are usually surrounded by elderly vendors selling their products without paying for a space within the market. Within the market, vendors include pensioners seeking to earn extra money, resellers who procure their goods from various places both abroad and nearby, and farmers and processors who have been marginalized from the formal retail sector or who prefer to sell at markets.

Within the confines of market territories, the vegetable, fruit, and honey vendors (who are permitted to sell outdoors) usually display their products on tables but they tend not to hang signs to identify their farm or business. Prices for individual items are sometimes displayed, as are handwritten labels that indicate the country of origin or that the produce is local. However, according to the consumers I interviewed, signs and labels cannot be trusted. Their concern centers around whether the vendor is actually a farmer (if they claim to be one), and whether the vendor can actually be trusted to sell quality, natural, and local produce. According to customers, the sign designating that something is local simply cannot be trusted; trust emerges instead through relationship building with vendors, whether they are resellers or farmers.

While buying from farmers is preferred and resellers tend to be stigmatized, the boundary between the legitimate farmer and the reseller is actually quite blurry. In informal interviews, many vendors admitted to buying and reselling for their rural neighbors who could not make it to the market, or buying from wholesalers to supplement their offerings. Resellers also claimed to offer a service to farmers who could not market their produce by themselves. Selling at the market demands time and money to pay for the space, in addition to the transport costs, a luxury that some small-scale farmers cannot afford. Ultimately, the low official entry barriers to selling at the market are conducive to some small-scale farmers but are also attractive to resellers and vendors. In fact, pensioners with allotment gardens, who crowd the entryways of the markets, are often in a position to offer the cheapest prices. This varied participation makes the market a competitive terrain in which resellers, farmers, and pensioners all offer prospective, if not guaranteed, possibilities for social and spatial embeddedness.

The story of what led one middle-aged vendor, Daiva, to sell at Kalvarijų Turgus helps elucidate one common trajectory for vendors and sheds light on the nature of the market interactions that may lead to the cultivation of social and spatial embeddedness (Blumberg 2014). Daiva and her family moved from the city to the country in the early 1990s and started farming on land restituted to her family. Like others, she explained, she thought that it would be possible to survive by starting a family farm, but these high hopes were met with continued disappointment. The latest problem her family farm confronted was the sharp decline in milk prices in 2008, which prompted her to get out of the dairy business and start selling honey in Vilnius. For Daiva, the financial crisis and its aftermath have not hampered her success in selling honey, and she even claims to have retained regular clientele. She explained that she has not had one complaint about her honey and that over time her customers have grown to trust her and to trust that her honey is local, pure, fresh, and unadulterated. When I ask if such trust is difficult to establish, she responds that of course, there are always lying vendors at the market, and that once she even had a vendor next to her who sold vegetables that were just "too clean" to be sourced directly from a farmer. Daiva's story demonstrates that at markets, social and spatial embeddedness is not guaranteed through face-to-face interactions, and that these interactions may even be defined as deceptive.

The interpersonal and face-to-face world of the public market does offer other advantages, if not a guarantee of social and spatial embeddedness. All sorts of possibilities for bargaining or striking a deal exist, especially when buying regularly or in large quantities. While customers may often pursue discounts, vendors sometimes proactively offer discounts. For instance, Daiva's flexible prices allow her to provide discounts to her less wealthy customers. Relationship making at the market is therefore a two-way process. To regular clientele, vendors offer discounts or free items by adding an additional vegetable or by rounding down on the scale to the customer's advantage. They provide trusted information, which is both lacking and desired in the markets; for example, they might make suggestions on the tastiest variety of a vegetable, or direct a customer toward the freshest items. They may also make small efforts to practice a form of social justice by providing discounts to those in need.

In summary, the trajectories that converge to make the place of the public market help explain why public markets continue to be significant everyday livelihood places for consumers, producers, and resellers, even though these markets are still regarded with suspicion by consumers. This type of simultaneous consumer skepticism and attraction to markets is the ongoing product of a trajectory of consumption that began in the Soviet period and coalesced in the 1990s, when consumers experienced markets as unruly but valued places. Later on, after the growth of supermarkets, public markets were not eradicated; in fact, supermarket growth produced the opposite effect: an appreciation by some consumers of the value and quality of the products at public markets. This perhaps surprising effect suggests that the making of place is also a relational process, as consumers ascribe value to certain places in relation to other places. Similarly, for the farmers involved, selling at public markets has become a livelihood strategy, in part because they have been marginalized from the conventional supply chains that provide supermarkets with produce. The dominance of small-scale farmers, a product of the reforms of the 1990s, has facilitated this process. Resellers who have been marginalized from the formal economic sector or who need to supplement low wages or pensions have also developed a livelihood strategy out of selling at public markets. Public markets are therefore heterogeneous places, where imported foods mingle with locally produced products, and where multiple actors are involved in building everyday livelihoods. Yet, they are also places *in process*: despite their well-entrenched appearances, the position of markets as privileged places for procuring local foods and for providing the possibility for social and spatial embeddedness has been challenged since the creation of new farmers' markets.

Trajectories Making Places: New Farmers' Markets

Already in 2007 small groups of farmers and consumers started experimenting with the creation of local systems of community supported agriculture, but in 2008, a conjuncture of events stirred more consumers and farmers to spearhead the creation of *farmers'* markets in Vilnius and other cities. While consumers were increasingly frustrated about the lack of availability of local produce in public markets, producers faced their own economic challenges (Melnikienė, Pelanienė, and Stončiuvienė 2013). A dramatic fall in milk procurement prices caused widespread and unprecedented protests by dairy farmers, and motivated some mid-scale dairy farmers to start selling milk directly to consumers, a practice that had been previously dominated by small-scale farmers. Policy-makers responded to this pressure by working with consumer and producer groups to create new and less rigorous regulations for the sale of local produce through alternative food networks.

Trajectories of consumption and production were therefore changing already before the onset of the financial crisis in late 2008. While the consequences of the financial crisis and subsequent austerity measures caused considerable societal distress (Woolfson 2010), they also heightened the economic importance of the local market both for consumers and producers. In other words, both social and spatial embeddedness became more coveted by farmers seeking markets and consumers seeking to support local farmers and obtain locally grown produce.

The idea of forging reconnection between farmers and consumers was an important motivation that spurred some residents of Užupis, a central Vilnius neighborhood, to organize a farmers' market called Tymo Turgus (or Užupio Turgus) in 2008. An organizer later recounted that at the time there was no other regular farmers' market, and that consumers were having difficulties procuring locally grown and organically certified food, while small-scale farmers were being increasingly marginalized from the conventional food market. But the initial impetus of organizing a farmers' market was not just to create a marketplace for local food, but to create a certain atmosphere: "Our goal is to collectively create a community market, which would attract those who want to live healthfully and openly. It would be simultaneously a meeting point and an event in which ecological products would be sold and tasted, where people would come and get to know each other, listen to music, stories, concerts, and enjoy life together" (Balsas.lt 2008). Originally planned for every other Thursday evening, the market opening hours were intended to accommodate working customers, who would arrive to shop after work and be greeted with music, concerts, and food, of course. Municipal authorities originally embraced the idea and allowed the use of Tymo Square, a newly built marketplace that had mostly stood empty, aside from occasional fairs and events (Verslo Žinios 2001). Organizers placed emphasis on the idea of trust and reconnection between consumers and producers, but in selecting farmers they first prioritized certified organic growers and they included other small-scale farmers and non-certified organic farmers to provide consumers with a greater diversity of products. By overseeing the market and allowing only certain farmers to participate in what was advertised as the Tymo Turgus of "environmental cultures," organizers sought to create a place that would guarantee both social and spatial embeddedness.

Several months following the inauguration of Tymo Turgus, other state and farmers' organizations started to publicly endorse the concept of a *farmers'* market. These organizations and institutions, including the Ministry of Agriculture, participated in a conference in December 2008, entitled *Promoting Lithuanian Farmers' Production*, which had the goal of encouraging mid- and small-scale family farmers to sell their home-grown produce at small markets. These farmers were in particular need of such assistance because they had been marginalized in public markets, which were allegedly dominated by resellers (ŽŪMI 2008). This initiative was also invested with a certain kind of meaning that emphasized the need to produce and consume natural, traditional, and Lithuanian food. For example, one government official highlighted the importance of encouraging "small- and mid-sized farmers to return to the forgotten natural, traditional production and selling of farm products" (ŽŪMI 2008). As announcements of the preliminary markets circulated, organizers also emphasized that quality and safety would be ensured for the natural, Lithuanian grown and made produce at the farmers' markets. Therefore, at the new farmers' markets, the natural and traditional would go hand-in-hand with the safe and modern.

After a few preliminary farmers' markets in Vilnius in December 2008, the initiative to organize markets was taken up by Lithuanian Farm Quality (*Lietuviško Ūkio Kokybė*), a cooperative organized by the Chamber of Agriculture of the Republic of Lithuania with the following mission: "the popularization of Lithuanian countryside and farm quality in cities, the strengthening of the connection between the city and the country, the popularization of natural food, the production of food that satisfies safety requirements for consumers, the encouragement of connections between the consumer and producer" (LRŽŪR 2011). Organizers determined that the best way to achieve these goals would be to create farmers' markets. However, instead of building a permanent marketplace that requires infrastructure, Lithuanian Farm Quality markets would be "mobile" and temporary, serving various city districts on different days of the week. Hence, the markets were officially called Mobilieji Ūkininkų Turgeliai (meaning small, mobile farmers' markets) or simply Mobilieji Turgeliai. While the temporary characteristic of the markets was borrowed from the farmers' markets in Ireland, which one organizer had witnessed, this characteristic also served to distance the new farmers' markets from existing public markets. Instead of being situated close to existing public markets, these markets mostly occupy spaces near supermarkets or shopping malls. But in contrast to supermarkets, on offer are traditional products, such as "natural milk, natural cottage cheese, natural farmers' cheeses, and also meat products, which are cold-smoked, traditional, and a heritage from the olden times" (Žilinskaitė 2009).

Like Tymo Turgus, the goal for the Mobilieji Turgeliai was to foster social and spatial embeddedness by providing a managed marketplace for selected farmers and local processors to sell their goods directly to consumers, but their strategies were remarkably different. While the original Tymo Turgus organizers were motivated to promote healthy and ecological living, Lithuanian Farm Quality created Mobilieji Turgeliai to reconnect consumers with an imagined Lithuanian heritage based in "natural" agriculture and processing, while at the same time advertising adherence to modern norms of food safety. They also had an ambitious goal to establish a weekly farmers' market in each of Vilnius' residential districts. Therefore, they sought to offer the possibility of social and spatial embeddedness on a more widespread basis.

Despite the differences, Tymo Turgus, the Mobilieji Turgeliai and other new farmers' markets were enthusiastically welcomed by Vilnius residents, many of whom became regular shoppers. However, the efforts to distinguish farmers' markets from public markets by enhancing the trustworthiness and credibility of the marketplaces has not prevented consumers from practicing the same type of skepticism and discerning behavior practiced at public markets. In other words, the culture that makes public markets has been reproduced at farmers' markets, despite efforts to ensure social and spatial embeddedness. In the following section, I describe the farmers' markets and I analyze the interactions that take place there, pointing out their continuities with the practices and trajectories that forge public markets.

Tymo Turgus

Tymo Turgus takes place on the outskirts of Vilnius' Old Town near the Užupis neighborhood, popularly considered to be a gentrified neighborhood and an artistic center for wealthier urbanites. Within walking distance of the central Cathedral Square and other popular tourist sites but far enough away from the bustling center to offer ample parking opportunities, the marketplace occupies an advantageous and also prestigious location. Unlike public markets near major roads and train stations, Tymo Turgus is not located near a major public transportation hub with its crowds and noise; in contrast, the atmosphere is even serene and peaceful. Situated downhill from the city's old defensive wall and the Vilnius barbican, Tymo Square is partially surrounded by a hill covered with scenic greenery. Its entrance is marked by a large wooden gate followed by identical wooden stalls lined up along a long wooden boardwalk. Usually the marketplace is quiet and empty, or occupied by a few groups of teenagers, but on Thursdays it experiences a dramatic awakening with the arrival of a bustling farmers' market.

On offer are fruits and vegetables, fresh and smoked meat and fish, honey, dairy products, baked goods, homemade jams, teas, and even plants and flowers. In the summer especially, farmers offer a greater variety of fresh vegetables than what is available at public markets, but otherwise the products are quite similar in appearance to those found at public markets. While some products like meat are usually wrapped and priced, typically there are not prices affixed to the boxes of vegetables. Some farmers hang their certificates, but others display no signs or certificates. When I inquired about this, farmers responded that there is no need for extra signs; their customers already know them. Despite its relatively brief history, Tymo Turgus appeared to be an established institution (Blumberg 2014).

The celebratory ambiance of the market persisted for its opening season, but the end of 2008 also marked the beginning of the financial crisis in Lithuania, popularly known as *sunkmetis* (hard times). The decline in purchasing power among the general population was felt in the market, and the municipality also started charging the market fees. The economic difficulties affecting the country are now evident in the physical structure of Tymo Square: repairs are badly needed to the boardwalk, which despite being relatively new, was not designed to withstand Lithuanian winters and has broken in several places. Musical and other events are now only rarely organized, and the market starts and ends earlier than was originally planned. Although the market was originally organized to include farmers who grow using organic production methods, the limited offerings of the organic sector meant that more conventional producers were eventually allowed to sell there too. In other words, the market transitioned from being a special event and a novelty centered on environmental living, to being a regular, weekly market, with hours that start in the morning and end in the afternoon.

Although the market's celebratory ambiance faded away and the market's physical infrastructure bears signs of wear, Tymo Turgus has endured as an institution and has even thrived. At first glance there is not much that distinguishes the products on offer at Tymo Turgus from those in other markets. The practices of bargaining and obtaining favorable deals are also present at Tymo Turgus. But absent are the vegetables and fruit that would be out of season in Lithuania, and no lines of elderly women clamor by the entrance with their goods. Prices also tend to be higher, and the clientele is dominated by mothers with children and wealthy working-age adults. Even a few celebrities shop at the market. Although vendors do not always post signs with their names, the market's website features photographs and descriptions of some of the vendors and their farms. The market's appearance and its advertising appear to enforce the notion that this should be a place where social and spatial embeddedness is guaranteed.

However, even this market has not been able to avoid a few instances in which unsanctioned vendors sold their goods, or permitted vendors resold produce that was not their own. These incidents did not surprise the customers I interviewed because they remain skeptical that building trust and reconnection could be completely delegated to the markets' administrators, the vendors themselves, or to certification schemes. Instead, even at Tymo Turgus, building trust requires building relationships, ideally to the point where you buy from your "own" farmer at the market. According to my interviewees, the most ideal method to ensure trust is to visit the farm itself to make sure that the farmer grows what he or she sells. Face-to-face interactions, although important, are clearly not enough to establish social embeddedness. Instead, social embeddedness is cultivated over time and becomes manifest in multiple ways. Customers exhibit their trust by returning to the same farmers and recommending them to others, and in return they may receive a little extra from the farmer or a more favorable reading of the scale.

For the marketplace as a whole, this kind of clientelization results in uneven patronage. Although over 50 farmers and producers may be present at any given time, long lines form at the stands of only a select few. According to a market organizer, competition has existed between market vendors, and suspicious rumors have circulated about the growing practices of more successful participants. Even though participation in the farmers' market does not in itself guarantee reconnection for farmers, gaining a place in the market is itself an achievement. A market organizer attested that she receives several phone calls from prospective vendors on a weekly basis, and unless they offer something different from what is already being sold, she has to turn them away. The trajectories that make this market are therefore limited by the restrictions designed to keep the marketplace differentiated from public markets by limiting the number and kind of vendors allowed to sell there.

Although these restrictions have an obvious impact on the appearance of the market (tropical fruits are not found here), they do not provide a guarantee of social and spatial embeddedness to shoppers. In everyday life, Tymo Turgus functions like public markets: customers emphasize building relationships, and even go so far as to visit farms. According to them, these kinds of efforts are necessary to establish trust and reconnection and to ensure that the products are actually locally produced.

Mobilieji Turgeliai

Like Tymo Turgus, the Mobilieji Turgeliai opened with much fanfare and press coverage, but unlike Tymo Turgus, which has not grown much since its inception, these markets have multiplied throughout Vilnius and in other cities as well. While only officially starting in early 2009, by January 2011 over 40 weekly markets were functioning in the most prominent residential districts of Vilnius and Kaunas, with about 200 participating vendors in total and 60 yearlong vendors. This spatial strategy was part of the organizers' effort to bring natural and local food to as many people as possible, but also to make sure that this food was affordable and adhered to quality standards. According to an organizer, by reselling foreign produce and claiming it as Lithuanian, vendors at public markets had tarnished the reputation of Lithuanian farmers. Therefore, part of the task of the Mobilieji Turgeliai was to regain consumer trust and confidence in the products of local farmers and certified traditional processors. The farmers and processors would in turn help ensure that trust by utilizing the logo of the cooperative.

The initial strategy organizers used to achieve affordability, quality, and accessibility was to help large-scale farmers restructure their businesses to include the capacity to process, deliver, and market their products directly to consumers. Producers who could take advantage of economies of scale were then able to quickly populate the weekly markets with their delivery trucks. Although not all farmers who sell at these markets operate on large scales, and the membership includes organic as well as conventional farmers, only a handful of farms and processors are able to sell at several markets during the week. The dominance of a few vendors can be explained by the capital investments required for participation. Because Mobilieji Turgeliai do not take place in permanent marketplaces with their own facilities, those selling fresh meat and dairy products need to invest in refrigerated trucks for transportation and for vending. While the necessary capital investments favored large-scale farmers, they also ensured that more neighborhoods would be served with markets more quickly.

In a typical market, vendors offer their products from refrigerated trucks, tables under tents, and sometimes also from the trunks of their cars or vans. Each market is organized so that there is a balance of dairy, meat, fish, breads, fruits, and vegetables, and the emphasis is on providing natural products. Some of the larger and centrally located markets attract more consumers and vendors, but most neighborhood markets have only four or five vendors at any given time. Markets continue to function during the winter despite the limited availability of fresh produce and lower numbers of vendors.

Consumer support of the markets is attested to by their growth and persistence, but already in the beginning there were problems controlling the marketplaces. The Mobilieji Turgeliai operate with low overhead costs, which means that there is actually very little oversight by organizers at the numerous individual markets. As a result, some of the more popular markets attract resellers as well as farmers who do not belong to the cooperative, but even some of the smaller markets have unaffiliated vendors set up tables or display their produce on the ground right beside the market. In effect, this exemplifies how the marketplace is porous and open, despite efforts to contain it and to publicize the market as a place only for certified vendors. The cooperative itself recommends that farmers advertise using the cooperative's logo and that the vendors wear aprons with the logo to distinguish themselves from intentional or unintentional interlopers (LRŽŪR 2011).

From the perspective of consumers, however, the logo itself does not provide a sufficient guarantee. According to the market practices pursued by consumers, caution is necessary when confronted with any labels, logos, or guarantees. As in public markets, consumers only establish trust after multiple transactions in which the purchased product has had consistent and good quality. The presence of resellers is not surprising for consumers, and resellers are not the only source of suspicion. Some consumers suspect that even bona fide farmers may import non-local meat and sell it as their own.

Despite the efforts of organizers of farmers' markets to carve out an exclusive marketplace for local farmers who produce and process using natural methods, consumers still approach these places with discerning and cautious behavior. When I asked customers why they shop at farmers' markets, sometimes they did not even recognize that they were shopping at a *farmers'* market; rather, they classified it as just another market or as a "little market" (*turgelis*). In summary, logos, advertising, and appeals to the essential Lithuanian qualities of the products available at the markets are insufficient to guarantee trust and reconnection, or social and spatial embeddedness.

Conclusion

Farmers' markets are increasingly capturing the imagination of policy-makers and academics in Western Europe and Northern America, where they are considered to be places of social and spatial embeddedness. They are places where face-to-face interactions produce trusting relationships and reconnection between consumers and producers of food (Kneafsey et al. 2008). For example, in his research on farmers' markets in Ireland, Kirwan found that "most consumers assume that the produce at FMs [farmers' markets] is somehow genuine, and in this sense the institution of FMs establishes a baseline trust for them" (2006, 307). Farmers' markets have also been documented as places that help make locally grown food more widely available and accessible. Gillespie et al. (2007) even argue that farmers' markets operate as a social economy by creating spaces that conjoin market transactions with social interactions, thereby enhancing civic life. However, the dominant literature on farmer's markets, which is focused on parts of Western Europe and Northern America, has not adequately accounted for how historical trajectories and contemporary practices forge spaces of consumption in interrelated and complex ways, thereby impacting how farmers' markets function to guarantee social and spatial embeddedness.

In this article, I utilize a comparative perspective based on a relational understanding of place to examine the potential of farmers' markets to foster social and spatial embeddedness in a post-Soviet context. As such, I make a contribution to the small but growing literature on alternative food networks and farmers' markets in post-socialist contexts (Jung, Klein, and Caldwell 2014; Spilková and Perlín 2013; Zagata 2012), but I also argue for the ongoing relevance of the scholarly literature on post-socialism and public markets, which helps make sense of what appear to be "new" or pan-European phenomena in post-Soviet contexts.

Vilnius is a city that already had a thriving market culture and persistent but evolving marketplaces before the introduction of farmers' markets. At public markets, clientelization helps farmers and consumers achieve social and spatial embeddedness, but this could also be a long and difficult process. Public markets are perceived with a contradictory perspective: they are valued for bargains, deals, and the potential for social and spatial embeddedness, but they are also despised for their unruliness and the potential for consumers to be cheated. The organizers of new farmers' markets promised to provide places that would ensure social and spatial embeddedness, but rather than becoming fundamentally different places, the practices that constitute new farmers' markets mirror the ones that make public markets. A relational understanding of place helps explain why these new places do not cultivate entirely new practices: similar trajectories of consumption and production forge a meeting point in both public markets and farmers' markets. This relationship has positively impacted the growth of farmers' markets because the practice of selling and buying at markets was already culturally intelligible, but it is a practice that does not resemble immediate and harmonious reconnection between consumers and producers. Even in the farmers' markets that strictly control the marketplaces, strive to ensure embedded transactions and cater to elite clientele, reconnection is not guaranteed. Nevertheless, the direct contact enabled by farmers' markets does facilitate relationship building, a source of consumer confidence that is more valued than labels.[5] Social and spatial forms of embeddedness are therefore possible but not simply guaranteed by the sociospatial context (farmers' markets) of market transactions. They are not unobtainable theoretical ideals, but an effect of the practices that form meeting points at farmers' *and* public markets.

In the European Union, local food networks are gaining increasing prominence in policy discussions at multiple scales. One of the proposed strategies to support the development of local food systems is to put in place a label specifically for local food (European Commission 2013). The analysis I have presented here casts doubt on the viability and suitability of such a strategy in Lithuania, where consumer skepticism and lack of trust in logos, brands, and certification schemes has been noted, and not only with respect to farmers' markets (Kavaliauske, Vaskiv, and Seimiene 2013). Indeed, this phenomenon has been documented and analyzed in different post-Soviet spaces, where distrust exists not only between vendors and consumers in markets, but also between society and the state (Matonytė 2006). Therefore, a new logo might simply become irrelevant, and the process of implementing and regulating a certified local food system may risk marginalizing those producers (and their consumers) who cannot meet the administrative requirements. I would like to conclude by suggesting that consumer distrust be viewed as a *strength* and a manifestation of resilience in coping with the ruptures and historical changes that have shaped Lithuania's post-Soviet and Europeanizing present (Mincyte 2012). Instead of introducing another label or another regulated marketplace, I suggest that attention must be placed on "vernacular" (Mincyte 2012, 41) forms of sustainability and how they already foster social and spatial embeddedness in distinct ways in different sociospatial contexts. Given the rich variety of alternative food networks in contemporary Lithuania, including informal milk delivery networks, nascent community supported agriculture schemes, and both farmers' and public markets, there are many places that already provide consumers and producers with the potential for reconnection and where vernacular forms of sustainability are practiced.

Acknowledgements

Thanks to anonymous reviewers for their comments, and to Ulrike Plath and Diana Mincyte for their encouragement and invaluable advice. All errors remain my own.

Disclosure statement

No potential conflict of interest was reported by the author.

Funding

This research was supported by the Association for the Advancement of Baltic Studies and the International Research & Exchanges Board. The Hella Mears Graduate Fellowship provided additional support to me to write this article.

Notes

1. I use the term "public market" or simply "market" to refer to bazaars, open-air and urban markets.
2. Because this research ended in 2013, recent changes that have occurred in Tymo Turgus are not part of this analysis.
3. See Mažeikis (2004) for a general discussion of markets in Lithuania.
4. See Klumbytė (2009) on changing consumption practices.
5. See also Skulskis et al. (2011) on organic products.

References

Aidis, R. 2003. "Officially Despised Yet Tolerated: Open-air Markets and Entrepreneurship in Post-Socialist Countries." *Post-Communist Economies* 15 (3): 461–473. doi:10.1080/1463137032000139106.

Aistara, G. A. 2014. "Latvia's Tomato Rebellion: Nested Environmental Justice and Returning Eco-Sociality in the Post-Socialist EU Countryside." *Journal of Baltic Studies* 45 (1): 105–130. doi:10.1080/01629778.2013.836831.

Alanen, I. 2004. "The Transformation of Agricultural Systems in the Baltic Countries – A Critique of the World Bank's Concept." In *Mapping The Rural Problem In The Baltic Countryside: Transition Processes in the Rural Areas Of Estonia, Latvia And Lithuania*, edited by I. Alanen, 5–58. Aldershot: Ashgate.

Alkon, A. H., and T. M. Mares. 2012. "Food Sovereignty in US Food Movements: Radical Visions and Neoliberal Constraints." *Agriculture and Human Values* 29 (3): 347–359. doi:10.1007/s10460-012-9356-z.

Balsas.lt. 2008. "Premjera: Ekologiškų Kultūrų Turgus Užupyje." Accessed March 9, 2014. http://www.balsas.lt/naujiena/202135/premjera-ekologisku-kulturu-turgus-uzupyje

Blumberg, R. 2014. "The Spatial Politics and Political Economy of Alternative Food Networks in Post-Soviet Latvia and Lithuania." PhD diss., University of Minnesota, Twin Cities.

Brown, A. 2002. "Farmers' Market Research 1940–2000: An Inventory and Review." *American Journal of Alternative Agriculture* 17 (4): 167–176.

Brown, C., and S. Miller. 2008. "The Impacts of Local Markets: A Review of Research on Farmers Markets and Community Supported Agriculture (CSA)." *American Journal of Agricultural Economics* 90 (5): 1296–1302. http://ajae.oxfordjournals.org/content/90/5/1298.extract

Czakó, Á., and E. Sik. 1999. "Characteristics and Origins of the Comecon Open-Air Market in Hungary." *International Journal of Urban and Regional Research* 23 (4): 715–737. doi:10.1111/1468-2427.00224.

DeSoucey, M. 2010. "Gastronationalism: Food Traditions and Authenticity Politics in the European Union." *American Sociological Review* 75 (3): 432–455. doi:10.1177/0003122410372226.

DuPuis, E. M., and D. Goodman. 2005. "Should We Go 'Home' to Eat?: Toward a Reflexive Politics of Localism." *Journal of Rural Studies* 21 (3): 359–371. doi:10.1016/j.jrurstud.2005.05.011.

Dzenovska, D. 2013. "The Great Departure: Rethinking National(ist) Common Sense." *Journal of Ethnic and Migration Studies* 39 (2): 201–218. doi:10.1080/1369183X.2013.723254.

European Commission. 2013. *Report from The Commission to the European Parliament and The Council*. Accessed March 9, 2014. http://ec.europa.eu/agriculture/quality/local-farming-direct-sales/pdf/com-report-12-2013_en.pdf

Feagan, R. B., and D. Morris. 2009. "Consumer Quest for Embeddedness: A Case Study of the Brantford Farmers' Market." *International Journal of Consumer Studies* 33 (3): 235–243. doi:10.1111/ijc.2009.33.issue-3.

Geertz, C. 1978. "The Bazaar Economy: Information and Search in Peasant Marketing." *The American Economic Review* 68 (2): 28–32.

Gillespie, G., D. L. Hilchey, C. C. Hinrichs, and G. Feenstra. 2007. "Farmers' Markets as Keystones in Rebuilding Local and Regional Food Systems." In *Remaking the North American Food System: Strategies for Sustainability*, edited by C. C. Hinrichs and T. A. Lyson, 65–83. Lincoln: University of Nebraska Press.

Goodman, D., E. M. DuPuis, and M. K. Goodman. 2012. *Alternative Food Networks: Knowledge, Practice, and Politics*. Abingdon: Routledge.

Hart, G. 2002. *Disabling Globalization: Places of Power in Post-Apartheid South Africa*. Berkeley: University of California Press.

Hohnen, P. 2003. *A Market Out of Place? Remaking Economic, Social and Symbolic Boundaries in Post-Communist Lithuania*. Oxford: Oxford University Press.

Holben, D. H. 2010. "Farmers' Markets: Fertile Ground for Optimizing Health." *Journal of the American Dietetic Association* 110 (3): 364–365. doi:10.1016/j.jada.2009.11.015.

Holloway, L., and M. Kneafsey. 2000. "Reading the Space of the Farmers' Market: A Case Study from the United Kingdom." *Sociologia Ruralis* 40 (3): 285–299. doi:10.1111/soru.2000.40.issue-3.

Hüwelmeier, G. 2013. "Postsocialist Bazaars: Diversity, Solidarity, and Conflict in the Marketplace." *Laboratorium: Russian Review of Social Research* 5 (1): 52–72.

Jung, Y., J. A. Klein, and M. L. Caldwell, eds. 2014. *Ethical Eating in the Postsocialist and Socialist World*. Berkeley: University of California Press.

Karner, S., ed. 2010. *Local Food Systems in Europe: Case Studies from Five Countries and What They Imply for Policy and Practice*. Graz: IFZ Graz.

Kavaliauske, M., U. Vaskiv, and E. Seimiene. 2013. "Consumers Perception of Lithuanian Eco-Label." *Economics and Management* 18 (4): 802–815. doi:10.5755/j01.em.18.4.4990.

Kirwan, J. 2006. "The Interpersonal World of Direct Marketing: Examining Conventions of Quality at UK Farmers' Markets." *Journal of Rural Studies* 22 (3): 301–312. doi:10.1016/j.jrurstud.2005.09.001.

Klumbytė, N. 2009. "The Geopolitics of Taste: The 'Euro' and 'Soviet' Sausage Industries in Lithuania." In *Food & Everyday Life in the Postsocialist World*, edited by M. Caldwell, 130–153. Bloomington: Indiana University Press.

Kneafsey, M., R. Cox, L. Holloway, E. Dowler, L. Venn, and H. Tuomainen. 2008. *Reconnecting Consumers, Producers and Food: Exploring Alternatives*. Oxford: Berg.

Knudsen, I. H. 2013. *New Lithuania in Old Hands: Effects and Outcomes of EUropeanization in Rural Lithuania*. London: Anthem Press.

LRŽŪR [Lietuvos Respublikos Žemės Ūkio Rūmai]. 2011. "Mobilieji Ūkininkų Turgeliai." Accessed March 9, 2014. http://www.zur.lt/index.php?1002735950

Massey, D. 2005. *For Space*. London: Sage.

Matonytė, I. 2006. "Why the Notion of Social Justice Is Quasi-Absent from the Public Discourse in Post-Communist Lithuania." *Journal of Baltic Studies* 37 (4): 388–411. doi:10.1080/01629770608629621.

Mažeikis, G. 2004. "Transparency and Functions of the Lithuanian Bazaar." *Lietuvos Etnologija: Socialinės Antropologijos ir Etnologijos Studijos* 4 (13): 49–66.

Melnikienė, R., N. Pelanienė, and N. Stončiuvienė, eds. 2013. *Lietuvos Žemės ir Maisto Ūkis 2012* [*Agriculture and Food Sector in Lithuania 2012*]. Vilnius: Lietuvos Agrarinės Ekonomikos Institutas.

Mikelionytė, D., I. Lukošiutė, and L. Petrauskaitė. 2010. "Promotion of Direct Marketing of Agricultural and Food Products in Compliance with Consumers Priority." *Management Theory and Studies for Rural Business and Infrastructure Development* 22 (3): 122–130.

Mincyte, D. 2011a. "Subsistence and Sustainability in Post-Industrial Europe: The Politics of Small-Scale Farming in Europeanising Lithuania." *Sociologia Ruralis* 51 (2): 101–118. doi:10.1111/soru.2011.51.issue-2.

Mincyte, D. 2011b. "Unusual Ingredients: Gastronationalism, Globalization, Technology, and Zeppelins in the Lithuanian Imagination." *Anthropology of East Europe Review* 29 (2): 1–21.

Mincyte, D. 2012. "How Milk Does the World Good: Vernacular Sustainability and Alternative Food Systems in Post-Socialist Europe." *Agriculture and Human Values* 29 (1): 41–52. doi:10.1007/s10460-011-9328-8.

Peck, J. 2005. "Economic Sociologies in Space." *Economic Geography* 81 (2): 129–175. doi:10.1111/j.1944-8287.2005.tb00263.x.

Polese, A., and A. Prigarin. 2013. "On the Persistence of Bazaars in the Newly Capitalist World: Reflections from Odessa." *Anthropology of East Europe Review* 31 (1): 110–136.

Sage, C. 2003. "Social Embeddedness and Relations of Regard: Alternative 'Good Food' Networks in South-West Ireland." *Journal of Rural Studies* 19 (1): 47–60. doi:10.1016/S0743-0167(02)00044-X.

Sage, C. 2007. "Trust in Markets: Economies of Regard and Spaces of Contestation in Alternative Food Networks." In *Street Entrepreneurs: People, Place and Politics in Local and Global Perspective*, edited by J. Cross and A. Morales, 147–163. Abingdon: Routledge.

Sik, E., and C. Wallace. 1999. "The Development of Open-Air Markets in East-Central Europe." *International Journal of Urban and Regional Research* 23 (4): 697–714. doi:10.1111/ijur.1999.23.issue-4.

Skulskis, V., V. Girgždienė, and R. Melnikienė. 2011. "Direct Marketing of Organic Products." *Management Theory and Studies for Rural Business and Infrastructure Development* 1 (25): 208–215.

Slocum, R. 2008. "Thinking Race Through Corporeal Feminist Theory: Divisions and Intimacies at the Minneapolis Farmers' Market." *Social & Cultural Geography* 9 (8): 849–869. doi:10.1080/14649360802441465.

Spilková, J., and R. Perlín. 2013. "Farmers' Markets in Czechia: Risks and Possibilities." *Journal of Rural Studies* 32 (October): 220–229. doi:10.1016/j.jrurstud.2013.07.001.

Verslo Žinios. 2001. "Tymo Kvartale Iškils Amatų Miestelis." Accessed March 9, 2014. http://archyvas.vz.lt/news.php?strid=1002&id=25494

Woolfson, C. 2010. "'Hard Times' in Lithuania: Crisis and 'Discourses of Discontent' in Post-Communist Society." *Ethnography* 11 (4): 487–514. doi:10.1177/1466138110372586.

Yükseker, D. 2007. "Shuttling Goods, Weaving Consumer Tastes: Informal Trade between Turkey and Russia." *International Journal of Urban and Regional Research* 31 (1): 60–72. doi:10.1111/ijur.2007.31.issue-1.

Zagata, L. 2012. "'We Want Farmers' Markets!' Case Study of Emerging Civic Food Networks in the Czech Republic." *International Journal of Sociology of Agriculture and Food* 19 (3): 347–364.

Žilinskaitė, R. 2009. "Mobilieji Ūkininkų Turgeliai Jau Veikia." Accessed March 9, 2014. http://verslas.delfi.lt/kaimas/mobilieji-ukininku-turgeliai-jau-veikia.d?id=20205773

ŽŪMI [Žemės Ūkio Ministerijos Informacija]. 2008. "Prekybos Tinklų Konkurentai – Natūralių Produktų Turgeliai." Accessed March 9, 2014. http://terra.zum.lt/lt2/naujienos/pranesimai-spaudai/6712/

CHANGING VALUES OF WILD BERRIES IN ESTONIAN HOUSEHOLDS: RECOLLECTIONS FROM AN ETHNOGRAPHIC ARCHIVE

Ester Bardone and Piret Pungas-Kohv

This article examines the historical importance of wild berries in the archival sources of the Estonian National Museum. The studied materials suggest that wild berries as food were insignificant for Estonian ethnologists-researchers as well as for correspondents due to disciplinary conventions and the ways of recollecting about food traditions. However, considering the Estonian remembrances in the context of international studies the consumption and gathering of wild berries for private use becomes a practice with diverse meanings. Wild fruits as food may have ambivalent values, which relate to socioeconomic factors, but likewise to continuities and discontinuities in individual and collective memory.

Today wild berries (botanically named "fruits") are considered an important feature of Estonian culinary heritage. The idea that berries have always been important part of the Estonian diet seems to rely on the fact that natural conditions in Estonia (e.g., climate, soil, and different habitats like forests and mires) are suitable for the growth of a variety of wild fruits. There are more than 20 species of edible fruits that grow in different types of forests and bogs throughout Estonia. The most common wild berries are European bilberry (*Vaccinium myrtillus*), lingonberry (*Vaccinium vitis-idaea*), wild raspberry (*Rubus idaeus*), wild strawberry (*Fragaria vesca*), bog bilberry (*Vaccinium uliginosum*), cloudberry (*Rubus chamaemorus*), and common cranberry (*Oxycoccus palustris*).[1] Estonian folk medicine has praised wild berries, among other edible plants, as remedies for various diseases (Kalle and Sõukand 2012).

Unlike the popular idea that wild berries have always been an important food for Estonians, a closer look at the history of wild berry consumption over the course of the late nineteenth and twentieth century reveals a more complicated picture where major historical events and the growing importance of transnational agricultural markets continued to transform local dietary cultures (cf. Lysaght 2000, 15). At the same time, people have different memories of the role of wild berry picking and consumption as part of the private food economy and everyday life practices. Our aim in this article is to examine how different generations of Estonians, throughout the last century and to the present, have recollected about the role of wild berries in domestic food consumption. Have they always considered wild berries a valuable food? How have people's memories about wild berries changed over time, and what has influenced these changes? These are the questions we want to address in order to reconsider today's popular belief that wild berries have always been an important food for Estonians.

Reflections on the Sources

To examine the ways in which wild berry consumption was remembered and how memories intersected with the historical changes, we primarily focus on ethnological questionnaires and their responses collected by the Estonian National Museum (ENM). The data stored in the archive of the ENM represent different political conditions defining three distinct periods in the development of ethnological research in Estonia, including the establishment of ethnology as a discipline in the Republic of Estonia (1918–1940), the development of ethnographic research during the Soviet occupation (1944–1991), and the reorientation toward the western European tradition of ethnographic research in the re-established Republic of Estonia (1991–present). Earliest recollections date back to the late nineteenth century, whereas the latest reach to 1990s. Such a time span enables us to see continuities as well as changes in wild berry memories of different generations of Estonians.[2]

As social historian Antoinette Burton suggests, the archive is a site of knowledge production and a mechanism for shaping the narratives of history (Burton 2006, 2). Drawing on the arguments developed by Burton (2006), our approach can be characterized as a retrospective interpretation of archival files. It is inspired by the scholarship concerned with the ways in which knowledge is shaped by the cultural environments, such as the work of cultural theorists (Assmann 2011; Manoff 2004), human geographers (Moore 2010; Ogborn 2010), and historians (King 2012; Dobson and Ziemann 2009; Steedman 2001). To examine correspondents' responses, we relied on the thematic analysis of archival responses (Riessmann 2008, 53–76). This included classification of the responses related to the issues of gathering and consumption of wild fruits in private households and correspondents' attitudes related to these practices. Additionally, we contextualized these themes with the data from other sources.

We selected four different surveys because (a) they specifically asked about the uses of wild berries in the households; (b) these questionnaires and correspondents' replies covered different periods in Estonian history; and (c) except for the study of responses collected in 2002 (Piiri 2006), other records have not been analyzed. More detailed information about the questionnaires, correspondents, and periods recollected can be found in Table 1.

The origins of ENM date back to the 1870–1880s, when Estonian intellectuals started collecting elder people's memories about traditional ways of life and folklore (for more on the origins and development of Estonian ethnology, see Kuutma and Jaago 2005; Annist and Kaaristo 2013). The German and Scandinavian examples of rescuing, collecting, and reconstructing traditional preindustrial peasant culture inspired the main principles for the establishment of the ENM in 1909. The main focus of ethnological research at that time was studying traditional material culture using the historic-geographic method.[3]

Ethnologist Ferdinand Linnus (1895–1942), following the example of Finnish and Swedish colleagues, decided to start the systematic collection of data with the help of local residents (e.g., teachers of village schools, farmers, craftsmen, etc.). He established a nationwide network of regular correspondents in 1931. The reasons for inviting people to contribute to the museum archive collections were mostly pragmatic: this was necessitated by the lack of researchers as well as the lack of funding for organizing extensive fieldwork trips all over the country. This way,

TABLE 1 Summary of sources from the Estonian National Museum

Year of compilation	1937	1947	1983	2002
Data about the questionnaires				
Reference	Linnus 1937	Sion 1947	Pärdi 1983	Piiri 2002
Number	No. 10	No. 43	No. 168	No. 214
Compiler	Ferdinand Linnus (1895–1942)	Virve Sion (1913–1985)	Heiki Pärdi (b. 1951)	Reet Piiri (b. 1955)
Theme	*Foods, drinks, flavorings*	*On picking mushrooms, berries, nuts, and other plant food*	*Gathering nature's gifts*	*Food culture in the Soviet period*
Total number of questions	95	16	22	174
Data about correspondents and the responses				
Correspondents' year of birth (the majority)	1880s–1890s	1890s–1900s	1910s–1920s	1920s–1930s
Volumes of responses in the archive (= KV)	KV No. 33, 50–52, 55	KV No. 77	KV No. 583	KV No. 1027–1033
Total number of replies	341*	36**	77	92
Time period recollected in responses***	1850s–1930s	1880s–1940s	1900s–1980s	1920s–2000s

Notes: * This is the total number of replies that included numerous responses from local teachers and schoolchildren who had collected information from their family members. The responses from the latter are not considered in our analysis.

** The correspondents' network had diminished considerably because some members had died in battle, some due to the repressions, others had fled.

*** Recollections also came from older inhabitants interviewed by correspondents.

correspondents-collaborators became co-creators of the collective memory of the nation.

The task of the correspondents was to prepare extensive written responses to the questionnaires and to interview elder inhabitants of the region (Tael 2006, 8). While current ethnological questionnaires encourage people to "freely write about their memories, experiences and values" and respondents are expected to express their opinions and attitudes, the instructions issued to the correspondents asked them to "objectively" represent a certain region and its culture, and, in so doing, help to collect the memory of the folk (Olsson 2011, 40–42; Värv 1990). Research aims and questions asked by ethnologists at different periods have determined what has been recollected and the methodology used (Kõresaar 1995; Jõesalu 2003). Throughout the twentieth century, questionnaires and correspondence was one of the most frequently used means for collecting data in Estonia as well as in Scandinavia, even if some changes occurred due to the changing political conditions disciplinary developments (cf. Hagström and Marander-Eklund 2005).[4] Because of Estonian ethnology's focus on the local folk culture and due to the Estonian origin of correspondents, the recollections gave little information about other ethnic groups (e.g., Baltic Germans or Russians).

The first questionnaire we choose under study was compiled in 1937 (KL No. 10). It addressed a diverse range of topics on peasant food culture from which only one question was related to the consumption of wild and garden berries. The responses also featured some earlier memories on how wild and cultivated berries were consumed as food. The structure and main topics of the 1947 (KL No. 43) and 1983 (KL No. 168) questionnaires were similar: the central content drew attention to the practices and customs of gathering berries, and only tangentially touched upon consumption.[5] The last, 2002 questionnaire (KL No. 214), provided a broader overview of the food culture of the Soviet period. Many questions focused explicitly on different ways of getting foodstuffs in the Soviet shortage economy and producing, preserving, and consuming food products at home.

In the following sections, we examine two major themes that emerged from the sources – memories about collecting wild berries and consuming them as food in private households. In the concluding reflections, we put Estonians' remembrances into a broader perspective bringing in comparisons from other European countries that share similar historical and/or environmental conditions.

Picking Wild Berries in Estonia: Berries as Supplementary Food

In order to understand how wild berries were remembered, people's memories should be contextualized with Estonia's socioeconomical background. Even though the Province of Estonia was part of Imperial Russia for most of its modern history (until 1918), it continued to be dominated by Baltic Germans who owned manorial estates and held high positions in the local administration. It is these elites who controlled the majority (80%) of the woodlands in Estonia's territory until the fall of the Russian Empire and the establishment of independent Estonia (Meikar and Etverk 2000, 8).

In their responses to the studies conducted between 1947 and 1983, correspondents remembered that in the nineteenth century the issue of ownership and forest maintenance had a significant impact on wild berry collecting. A farm wife (b. 1897) recollects: "picking berries and mushrooms in manors and state forests was prohibited, especially from young forests, because pickers trampled on the young trees" (KV 43, p. 22).

For Estonian peasants, berry picking was associated with the restricted access to forests. The majority of the correspondents mentioned the need for permission to collect berries and the symbolic payment (consisting of a part of the harvest, or also a small tax) to the manor. The most frightening stories were recalled about rangers (often of Estonian origin) who mistreated the pickers. The manors' restrictions on gathering had two major reasons. The Russian Empire introduced the forest preservation law in 1888 that required more careful maintenance of private forests in its provinces (Teplyakov 1998, 7). Additionally, encouraged by the Russian gentry, Baltic German landlords started seeing wild berries as a resource for additional income (Jürgenson 2005, 56–57). Correspondents also mentioned berry export by traders, who transported them to Riga and St. Petersburg, as a factor for the increase in commercial use of wild berries.

In the late nineteenth to early twentieth century, consuming berries and valuing their taste and nutritious qualities were mostly associated with the higher social classes, such as the Baltic German gentry, wealthy citizens, and the emerging Estonian social elite, whereas gathering berries was related to the lower classes, mostly poor landless people (e.g., cottagers, freedmen) who picked them mostly to sell.

Recollections about berry picking from this era are saturated with memories about market exchanges. For instance, a farm laborer (b. 1916) remembers: "[W]e sold them to manor gents on the spot, even the baroness and the doctor's wife at Kose bought some. City gents and ladies paid good money, 10 kopecks for a can [of wild strawberries]... If you were a careful picker, you could make up to 20–30 rubles in this way" (KV 77, p. 59).

At the same time as peasants sold berries to the local elites, their own diets were dominated by food produced on farms. The main pragmatic reason for limited berry consumption by Estonian peasants was the fact that the ripening of the berries coincided with more important summer and autumn tasks on the farm. A farm worker describes that bilberries, cloudberries mature in mid-summer during haymaking season, lingonberries during oat collection period, bilberries at the same time as processing rye; and cranberries during potato harvests (KV 77, p. 76). Although Estonians also picked berries for household use, this was done on Sundays or after the important farm tasks were completed. Therefore, among the social groups, children and old people (primarily women) were often mentioned as the main berry-pickers because they were not fully employed in farmwork. In some responses, city folk were reported to travel to the countryside to pick berries for their own consumption as they had more time and less opportunity for food production.

After the newly established government of the Republic of Estonia passed the Land Act in 1919, land ownership was reorganized and landless people, among which

Estonian-speaking peasants constituted a majority, became proprietors. As a result of agricultural reforms, around 80% of the forests remained state owned and managed (Meikar and Etverk 2000, 13). According to Estonia's laws, berry picking from state forests that were older than 15 years was free, and gathering restrictions applied only to young stands (Kusmin and Kusmin 2011, 19). In the 1930s, Estonia was predominantly an agricultural society and the forested land comprised only about 20% of the territory (Etverk and Sein 1995, 405).

During this time, berries constituted a considerable source of additional income for the poorer rural and urban inhabitants in the 1920s–1930s. In addition to local markets, berries were also sold abroad: in the 1930s, Estonia exported wild berries such as cranberries and lingonberries to 20 countries; at the period, berry picking was a considerable source of income for approximately 3000–5000 people in Estonia (Paal 2011, 68). Because most of the commercial berry pickers were women, the common word used for such pickers in the press of that time was "berry-women".

Furthermore, berry picking was significantly impacted by the introduction of new horticulture in the course of the twentieth century. Responses to the 1937 questionnaire revealed that in the nineteenth century berries were mostly eaten by "people of rank". In the first decades of the twentieth century, growing fruit trees and berry cultures became more widespread. Having acquired the plants and knowledge about domesticated berries, the land-owning Estonian farmers began to cultivate them in their estates (Banner 2005, 307–311). Thus, during the years of independence, the need to collect berries in the wild became less necessary.

After Estonia became part of the Soviet Union in 1944, all the land was nationalized and the forests became the property of the Soviet state. The newly established collective and state farms relied were notoriously poorly equipped and staffed by unmotivated workers, and overall production decreased in comparison with the prewar period (Klesment 2009, 251). Due to such organization of production, food shortages were a common part of everyday life in Soviet Estonia in the 1960s and 1970s (Raun 2001, 198–203). In the shortage economy, both rural and urban inhabitants had to look for additional food sources by cultivating food on subsidiary farms and kitchen gardens, especially from the 1940s to the early 1960s, when collective farms did not pay sufficient monetary wages (Klesment 2009, 257). However, the private household plots were too small, only 0.5–0.6 ha (Piiri 2006, 57–58), to supply food for the entire household.

It is in this context that economic values of wild berry picking became particularly important. Not only did individual pickers entered forests, collective berry picking was also organized by the state and workplaces. During the season, public transportation made it possible to access to the berry habitats easy and teams of pickers from a particular workplace went to the forests together. In her recollection, a teacher (b. 1925) describes: "Cranberry picking became particularly fashionable in those years [1960s–1980s]. Especially when a specific date had been set, there were vehicles: buses, trucks and cars… The marshes were full of machines… It was a 'cranberry fest'–like a large market" (KV 583, p. 246).

This way picking berries became an activity that enforced a feeling of community and created collective identities under socialism (Piiri 2006, 64). The picking season was regulated by the state to protect berry habitats, but the results of such restrictions were rarely followed, and setting the dates was later replaced with new forms of supervision and forest management.

Additionally, because the Soviet Union could acquire foreign currency from the sale of berries, berry collection and stocking standards were established in the 1960s–1970s to meet international market expectations. While there are no data about how much of the wild berries were collected for sale and for private use, the responses from the ethnological studies suggest the number of berry pickers swelled during the 1970s–1980s to the point that the undergrowth of many mires were trampled (e.g., in some cranberry bogs up to 95% of the biological harvest was picked) (Paal 2011, 70). Some correspondents criticized the unsustainable management of natural resources in the Soviet Union that had resulted in irreversible changes to berry habitats. For example, a retired gardener (b. 1919) condemns that:

> There were bilberries, lingonberries and cranberries. It was that way about a decade after the war. But then everything began to change in nature... Forest cutting left large areas bare and the bilberries perished... Marshes and bogs that once grew cranberries quite plentifully, are now dried out because of the melioration and cranberries have disappeared completely. (KV 583, p. 106)

This recollection refers to the negative image of the Soviet management of natural environments, especially wetlands, which were treated as economic resources to be exploited on a large scale. In reaction to the intensive economic pressure, the environmental protection movement became active in the early 1980s, and as a result, protection areas were designated in Estonian wetlands that were considered important for hydrology and for the profusion of berries (Aber, Pavri, and Aber 2012, 257).

In 1983, correspondents were likewise asked about the importance of wild food in the diets of their families. The majority of correspondents declared that the wild food was of medium importance in their family diet but could potentially be more significant. As a librarian (b. 1921) writes: "I especially like mushrooms and wild berries, [and] their importance as a family food could be bigger. Gathering them is limited by the lack of time. The importance of home gardens has increased so that picking gifts of nature constantly decreases. Who does not have a private garden uses more gifts of nature" (KV 583, p. 373). The gatherers had to find additional time for foraging alongside paid labor and taking care of private subsidiary holdings. Therefore, garden berries and fruits from one's own garden plot were more important for the daily diets due to easy accessibility, even if the taste of the forest bounty was highly appreciated.

The 1983 questionnaire KL No. 168 (compiled by Pärdi 1983) echoes earlier responses that emphasize wild berries as a desirable food, but it also points to the growing interest in berry picking as a leisure activity: "[T]oday gathering gifts of nature is a necessity only for a few. Nevertheless, most of the people cannot imagine their life today without wild berries, mushrooms or nuts" (KL No. 168). There were new values that wild berries were assumed to have within a modern and increasingly more urbanized society such as "gathering as a pleasant alternative" and "an opportunity to be in nature" (KL No. 168). In her response, a collective farm worker (b. 1921) explicitly referred to the change in values of berry picking: "[T]oday picking gifts of nature is no longer vital; it is more like a former habit to enrich the table. It is useful and pleasant for everyone to linger in nature, especially for the city folk" (KV 538, 317). This suggests that in the late socialist era a novel recreational value was associated with gathering berries, especially by urbanites who formed the majority of the correspondents by the 1980s.

Wild Berry Consumption in Estonian Households: From the Raw to the Preserved

At the end of nineteenth century, wild berries provided a small seasonal supplement in summer and autumn, when meat stocks were running out, or because of other causes of duress, and primarily for poorer families (see also Luts 2008, 131–132; Kalle and Sõukand 2012, 278–279). Once agriculture was the main source of living in the late nineteenth century, the Estonian peasants' diet was rather unvaried, and its main components were rye bread, potatoes, fish, pork, milk, and different cereals (Moora 2007, 35). Since the 1880s, nutritional advice had been given in the press as well as in cookbooks, and from 1910 domestic economy and gardening courses became organized by different societies that aimed to improve Estonians' knowledge of food production and consumption. During the 1920s–1930s, multiple cooking courses and domestic economy schools stressed the importance of a varied diet and spread the know-how to Estonian farmwives (Troska and Viires 2008, 277). Such developments had an impact on how fruits, including wild berries, came to be used in the everyday food culture of Estonian farms.

The 1937 questionnaire (KL No. 10) asked correspondents: "[D]id wild or garden berries have any importance in the food economy during olden times?" Most respondents stated that in "olden times" berries were eaten raw and were consumed as a seasonal sweet. Thus, early recollections, dating back to the last decades of the nineteenth century, brought up a classical question for food researchers: the distinction made by Claude Lévi-Strauss ([1965] 1966) between *the raw* and *the cooked*. In light of this distinction, how did the experience and status of berry consumption change during the twentieth century? What were the reasons for this change? These are the questions we would like to address next.

As in earlier studies, the majority of correspondents involved in the 1937 survey were men who rarely cooked. Nevertheless, there were some quite detailed responses about what kind of berries were eaten in farms and how. For instance, a farmer (b. 1877) describes:

> Wild strawberries and bilberries were eaten while ripened, with fresh milk. Other berries, such as raspberries [and] stone brambles, as well as garden berries such as gooseberries and black and red currants were eaten as they were. After they had ripened, no desserts or wine were made [of these berries] … Formerly sugar was expensive … Berries were just for sweets, and so were apples too. Rowanberries[6] were brought home for winter and if they had been touched by the

cold they tasted sweet. Cranberries were likewise eaten as they were in winter, later they and other berries were used for making all kinds of desserts like *kissell* [fruit drink thickened with potato starch], compote, jam, marmalade, wine and others. (KV 33, p. 108)

This description gives quite a good overview of the ways that wild as well as garden berries were consumed in Estonian rural households, although it is not specified when the practice of preservation and cooking berry dishes became more common. From the other sources, it became clear that the major changes in using more varied fruit dishes took place at the turn of the century although these remained festive food until the early 1930s, when peasants' living standards and farmwives' knowledge improved considerably.

The limited consumption of berries had practical socioeconomic reasons, among them the high cost of sugar mentioned by the correspondent. The lack of sugar limited what berries could be consumed, and how.[7] From the perspective of modern science, we can say that the biochemical properties such as the high concentration of organic acids worked as natural preservatives that enabled the preservation of raw cranberries and lingonberries over winter, without added sugar. Until the twentieth century, the means and technologies for cooking and preserving were rather rudimentary (e.g., the lack of containers). For instance, special stoves for cooking connected to the chimney and separate year-round kitchens became extensively built in Estonian farmhouses only from the 1880s when peasants' living conditions started to improve (L'Heureux 2010, 480; Pärdi 2012, 53). It was only at the turn of the century when additional tools for cooking and preserving came into use (Moora 2007, 55–56).

In the early decades of the twentieth century, the accessibility of sugar depended on social status, and making preserves was limited to those who could afford sugar. According to the correspondents' memories, the availability of sugar was limited both by the food crisis during World War I and then later by its high price. For these reasons, berry preserves were valued in correspondents' responses as a special food eaten on special occasions, offered for guests or consumed as medicine (e.g., raspberry jam given to children against the common cold). The Estonian farmers began to pick berries more extensively for domestic use when living standards improved and the accompanying possibility of buying sugar boosted home preserving. According to the data collected by a correspondent (b. 1916): after the First World War, some families turned 5–6 poods[8] of sugar into jam every summer, other richer folk made a lot more (KV 77, p. 45).

In terms of methods of preserving, the correspondents indicated that at the end of the nineteenth and in the early decades of the twentieth century, the preserves were not kept in the small glass jars widely used today, but in clay pots or tubs made from alder or spruce stored in cellars or potato cells. Glass jars became more widely used in farmsteads during the 1920s–1930s, but the means for sealing them were still rudimentary (e.g., parchment or cellophane).

One more important factor promoting the culture of preservation and the knowledge about preservation was briefly mentioned in correspondents' recollections. In her response, a former farmer exemplifies: "[E]arlier (about 75 years ago – [in 1870s]) picking berries for one's own use was unknown. But at least 40 years ago [approximately since 1907] they already knew how to preserve and use various wild berries" (KV 77, pp. 129–130).

How such knowledge was distributed was specified only in a few responses from 1937 and 1947. Two groups were named as mediators of the knowledge related to conserving: the Baltic German gentry and the urban inhabitants, mostly Estonians who worked in a manor or had moved into cities. Correspondents attributed the spread of preserving know-how mainly to personal contacts and did not mention topics discussed in the educational literature of the time (e.g., berries as diversifiers of diet and their nutritional value). Although the knowledge about preserving and cooking berries (e.g., making berry preserves, wines, baking cakes, etc.) in order to diversify the diet and make it more nutritious was communicated to the Estonian public already in the late nineteenth century in gardening manuals (e.g., Spuhl-Rotalia 1898), the practice of making preserves spread much later. It became common among rural Estonians only when living standards on farms improved, sugar became more affordable, and women in farms had more time to deal with conserving (Moora 1981, 568).

In the Soviet shortage economy, self-made preserves were part of people's everyday life. The 2002 questionnaire (KL 214) addressed the topic of wild berries among numerous questions about preservation and consumption of the fruits, and, therefore, recollections focused on these issues were better represented. Several correspondents saw making preserves in the Soviet era as a continuation of the self-sufficiency economy stemming from the prewar Republic of Estonia. According to the remembrances collected by a female librarian (b. 1945):

> In the countryside, as a consequence of farm-keeping, it was clear that every farm must make its own preserves. It was unthinkable that for example in the 1950s someone living in the countryside would buy such things from the store. This habit and custom lasted for a long time. Later it was realized that self-made preserves were cheaper and also tastier. (KV 1031, p. 220)

To assess the proportions of garden and wild berries, the survey "Personal Subsidiary Holdings" conducted in 1978 demonstrated that 45% of all families made jam, 31% pressed juice, and 28% made compotes. In general, approximately 80% of all produce from personal households was turned into preserves (Raig 1981, 39). This suggests that there was a general increase in berry and fruit preserves in people's diet in these decades. Preserves were used in a variety of ways and became a common part of the food used during festivities, but also in everyday contexts.

However, in spite of the need and interest in making home preserves, the means for that were not always available. Several correspondents remembered the sugar crisis in the 1940s and referred to the ration cards that were used in the 1950s to distribute many foodstuffs. Therefore, wild berry species that required little or no sugar and entailed traditional ways of preserving remained important in these decades. Preserve making started to spread faster in the 1960s when the sugar crisis abated and metal jar lids and jar lid fasteners appeared on the market, guaranteeing an air-tight seal (Piiri 2006, 67).

It should be noted some correspondents saw that preserve making not simply as a necessity in the shortage economy, but as a creative challenge involving competitiveness. For instance, a librarian (b. 1937) wrote that in her family the cloudberry compote was considered the best: "One year I made 76 liters of cloudberry compote. When I told this to others at work, they thought I was a liar" (KV 1031, p. 62). This suggests that making preserves was a culturally and personally important endeavor. In a similar way as berry picking, it was a way of spending time with the family, creating continuity between generations, and a method for self-realization (Piiri 2006, 77–79).

Finally, a specific difference between Soviet and Western preservation habits must be mentioned. If in western Europe deep-freezing was already common since the 1960s (Shephard 2006), this technology was unavailable for most individual households in Soviet Estonia, where refrigerators with deep-freeze chambers were only sold from 1975 onward (Eelmaa 1985, 5). In their responses, correspondents mentioned that it was only in the post-Soviet era, in the mid-1990s, when refrigerators with more spacious deep-freezing chambers and freezers became available in stores. Due to this, making berries into jams was partially replaced by refrigeration that saved time as well as vitamins and nutrients and made the whole process of preserving much easier.

Concluding Reflections: Estonians' Wild Berry Memories in International Context

In his analysis of the development of ethnological research in Sweden, Fredrik Skott characterizes the collection of folklore archives as a project of producing collective memory and communal identity (Skott 2008, 282). Similarly to Scandinavian tradition, for correspondents in Estonia, participation in ethnological research and contribution to ethnographic archives became a way to write local histories and to contribute to the creation of collective memory. Additionally, the studied sources provided an interesting data for a contemporary researcher to see how and what was recollected about cultural phenomena at different times.

Due to the principles dominating ethnological research, the recollections from 1937, 1947, and partly also from 1983 reflected impersonal descriptions collected from elder inhabitants. People's personal stories about how they collected and consumed berries and why they valued them were often left out or kept in the background because of the way questions were formulated (e.g., short answer questions or multiple-choice answers) and research scope (e.g., focus on describing practices rather collecting individual memories). However, even fragmentary recollections enabled to draw connections between different periods. More detailed personal remembrances on food culture and the consumption of wild berries could be found from biographically contextualized responses collected in 2002 that revealed not only collective but more individual meanings and values that berry picking and eating has.

In the following reflections, we compare some of the main results and conclusions based on our sources with international studies in order to put Estonians' remembrances into a broader perspective indicating to cultural, political, and economic dimensions that have shaped how people have valued wild fruit consumption.

Correspondents' reports that wild berries constituted an important source of additional income and nutritional value for poorer families are supported by similar data from other countries (cf. Hietala 2003; Bringéus 2000). Ken Albala argues one's relation to wild food highlights social and cultural issues, among them different social class relationships to nature (Albala 2006, 9–18). Our study likewise showed that social status intersected with both the collection and eating of berries in significant ways. Recollections dating back to the end of nineteenth century, early twentieth century showed that Baltic German gentry and the upper classes in Estonia valued the taste of various kinds of fruits and had sugar as well as knowledge for preservation that were unavailable to peasants (cf. Notaker 2003, 559). The limited access to sugar restricted the berry species that were collected for household use; therefore, Estonian peasants, like their counterparts in European countries, consumed bilberries, lingonberries, and cranberries that contained organic acids and could be preserved without sugar (cf. Łuczaj et al. 2012, 362; Svanberg 2012, 319–320). More importantly, there was a deep social gulf between peasants who picked berries and higher social classes who consumed them.

During the first decades of the twentieth century when Estonia was mainly an agricultural society, farm produce became increasingly more valued than wild food; therefore, correspondents considered wild berries rather marginal in the national diet. Prioritizing farm production was also the reason why berry pickers were typically those who could not be fully engaged in farm work, including landless folk before 1919 and especially women (cf. Svanberg 2012; Lindquist 2009; Łuczaj 2008). In a similar way, time and energy invested in individual food production in Soviet Estonia made many correspondents value agricultural and horticultural produce over the consumption of wild berries that were proportionately less consumed. In the Soviet shortage economy that relied on self-provisioning where most of the food was produced in subsidiary holdings and gardens, foraging for food from nature played an important role (cf. Bellows 2004; Caldwell 2004). In contrast to their Soviet counterparts, Scandinavians considered wild berry picking as part of their summer cottage culture and a leisure activity (Pouta, Sievänen, and Neuvonen 2006; Lindhagen and Hörnsten 2000). For Estonians, berry picking as a recreational practice emerged only in the late Soviet period when the country became increasingly more urbanized and the standards of living improved.

We also found from correspondents' recollections that during different political regimes wild berries remained an important export product that facilitated commercial harvesting of berries in the region throughout the twentieth century and guaranteed Estonians extra income (cf. similar results in Sweden and Finland in Bringéus 2000, 183–186; Hietala 2003, 186–193). Berries were likewise sold at the local market or exchanged for goods. In Soviet Estonia, the intensified collection of wild berries had several socioeconomic components, including state-organized berry picking campaigns that made it into a site for building collective bonds. The shortcomings in state forest management resulted in prioritizing the quantity of berries picked over the damage caused to the environment. Thus, similarly to neighboring countries Estonians' berry picking memories were influenced by the emergence of global berry markets, the increasing industrialization of agricultural production, changes in local horticulture that included the introduction of domesticated fruit and berry plants, the modernization of rural households, and the proliferation of new technologies. However, in Scandinavia, the popularity of harvesting wild products has been related to the so-called Everyman's Right (the right to forage for wild berries and mushrooms also in private lands) that emphasizes cultural values. For instance, Estonians' neighbors, the Finns, understood berry picking not just in terms of pragmatic needs, but also as symbolic values, including a sense of belonging to nature (Pouta, Sievänen, and Neuvonen 2006, 288–289; Hietala 2003, 193–194).

Another conclusion drawn from our sources is that correspondents did not consider wild berries as a proper food in comparison with the farm produce and cooked food. This does not amount to saying that the lower social classes did not eat wild berries, or that that their nutritional value and status in comparison with the farm produce was considerably lower. Rather, wild berries were valued as a complementary or healing food alongside other herbal remedies (cf. Hietala 2003; Łuczaj et al. 2012; Svanberg 2012). The status of wild berries as food in correspondents' recollections varied along with the changing social status of the consumers; changing nutritional and culinary knowledge; the improvement of home cooks' kitchen literacy; and the development of new cooking technologies.

The accessibility of sugar, an improvement in cooking technologies, the means for preservation, and the wider distribution of knowledge of how to make conserves from the 1920s on were the key factors that increased and varied the domestic consumption of wild as well as garden berries, and changed their food status. Making berry preserves boosted in the 1960s–1980s when the additional means and technology made it possible, and various fruit and berry conserves became a considerable part of everyday food consumption (cf. Łuczaj et al. 2012, 362).

The continuity of the peasant food economy can also be detected in the Soviet preservation practice. Like in other countries belonging to the Soviet bloc, making preserves was perceived as not just a labor-intensive duty, but also as an annual family ritual and a form of creative self-expression (Caldwell 2004, 100–114). In post-socialist period, deep-freezing marked the transition from the preserves-centered berry economy into the new era of wild food consumption in Estonia.

To return to the questions raised at the beginning of the article, meanings and values attached to wild berries changed significantly from 1937 to 2002. Berries did not always evoke pleasant nostalgic memories, like it may happen today, but were related to the ambivalence with which people engaged in foraging and poverty-driven consumption of wild foods. In the recollections of preindustrial age, wild berries were remembered to be collected for pragmatic reasons. In the context of modernization and urbanization, however, the correspondents emphasized the noninstrumental value of the wild bounty. Social distinctions as well as existing food hierarchies have influenced limited consumption of wild berries in different times. Today, foraging is increasingly given a recreational value and is associated more with personally significant memories and meanings rather than with the necessity to sustain oneself (Bardone 2013, 38–41). Besides, picking or eating wild berries may have different importance also in the same person's life at different periods: metaphorically speaking, the taste of wild fruits in the childhood may not be the same in the old age. But this is a topic for a new study.

Acknowledgments

We are grateful to Diana Mincyte, Ulrike Plath, Ene Kõresaar, Kirsti Jõesalu, Susanne Österlund-Pötzsch, Yrsa Lindquist, and two anonymous reviewers for their valuable suggestions. We also thank Tiina Tael and Reet Ruusmann from the Estonian National Museum for their kind assistance.

Disclosure Statement

No potential conflict of interest was reported by the authors.

Funding

This research has been supported by the European Union through the European Regional Development Fund (Centre of Excellence in Cultural Theory); the Estonian Ministry of Education target-financed projects [SF0180049s09] and [SF0180157s08]; the institutional research funding [IUT3-2] and [IUT 34-32]; and the Estonian Science Foundation [grant numbers 8040 and 9419].

Notes

1. Because wild fruit plants have multiple parallel names in English, we added Latin names in order to facilitate the identification of the species. The Estonian and Latin names are correlated according to the Index of Estonian Plant Names (see Eestikeelsete Taimenimede Andmebaas 2014) and to the international database The Plant List (see The Plant List). Because Latin plant names also have synonyms that knows a plant, it is necessary to clarify that *Oxycoccus palustris* is a Latin equivalent for the *Vaccinium oxycoccos*; the former name is more commonly used in Estonia for the common cranberry.
2. The majority of correspondents were of Estonian ethnic origin and only a few had a mixed Russian-Estonian ethnic background in 1983. In the 1930s, more than half of the correspondents were teachers, with farmers, artisans, officials, and others making up the rest. In 1937 and 1947, the majority of correspondents were rural inhabitants. In 1983, correspondents lived both in rural and urban areas, and in 2002 mainly in urban settings, having diverse occupational backgrounds. Prior to World War II, men dominated among the correspondents, whereas after the war mostly women contributed to the archive. In 1939, the correspondents' network had 413 members and from the 1950s to the 1980s there were around 200 active members comprised mainly of retired people.
3. The historic-geographic method was developed in the 1870s by Finnish folklorists Julius and Kaarle Krohn and was adapted to ethnology in order to compare the regional variations of ethnographic phenomena and their changes in time. The method was especially useful for investigating details, similarities, and differences in material cultures (Goldberg 1984).
4. There have been other methods of collecting data in Estonian ethnology, including ethnographic fieldwork, biographical methods, and others.
5. Similarly, a Finnish questionnaire from 1970, *Metsämarjat* [Wild berries], includes 10 thematic blocks of questions, of which only one is related to consuming wild berries

as food (the compiler was Riitta Ailonen; for the analysis of the responses, see Ailonen 1977).

6. The Latin names of the berries mentioned in the quote are *Rubus saxatalis* for stone bramble and *Sorbus aucuparia* for rowanberries. The former grows mainly in the coastal areas and islands, whereas the latter is common in all Estonia.
7. Drying, as a means for preserving berries, was mentioned only by a few correspondents, and this primarily for the purpose of making medicinal tea (cf. Kalle and Sõukand 2012). Adding dried berries to bread dough, or making berry-porridge was not remembered by the correspondents, although in eastern regions of Estonia berries were used this way (Moora 1981, pp. 568–569).
8. "Pood" was a unit of measurement used in Imperial Russia. One pood is approximately 1638 kg.

References

Ethnological Sources

Ethnological questionnaires (No. 10, 43, 168, 214) and the responses in the Correspondents Archive (KV No. 33, 50–52, 55, 77, 583, 1027–1033).

Printed Sources

Aber, J. S., F. Pavri, and S. Aber. 2012. *Wetland Environments: A Global Perspective*. Chichester: Wiley-Blackwell.

Ailonen, R. 1977. "Luonnonvarasten Marjojen Poiminta ja Käyttö." Masters' thesis defended, Helsinki University.

Albala, K. 2006. "Wild Food: The Call of the Domestic." In *Wild Food. Proceedings of the Oxford Symposium on Food and Cookery 2004*, edited by R. Hosking, 9–19. Devon: Prospect Books.

Annist, A., and M. Kaaristo. 2013. "Studying Home Fields: Encounters of Ethnology and Anthropology in Estonia." *Journal of Baltic Studies* 44 (2): 121–151. doi:10.1080/01629778.2013.775846.

Assmann, A. 2011. *Cultural Memory and Western Civilization: Functions, Media, Archives*. Cambridge: Cambridge University Press.

Banner, A. 2005. "Aeg, Aed ja Inimene. Aianduse Tähtsusest Eesti Maamajapidamistes." In *Eesti Looduskultuur*, edited by T. Maran and K. Tüür, 305–325. Tartu: Eesti Kirjandusmuuseum.

Bardone, E. 2013. "Strawberry Fields Forever? Foraging for the Changing Meaning of Wild Berries in Estonian Food Culture." *Ethnologia Europaea* 43 (2): 30–46.

Bellows, A. C. 2004. "One Hundred Years of Allotment Gardens in Poland." *Food & Foodways* 12 (4): 247–276. doi:10.1080/07409710490893793.

Berendsen, V., and M. Maiste. 1999. *Esimene Ülevenemaaline Rahvaloendus Tartus 28. Jaanuaril 1897*. Tartu: Eesti Ajalooarhiiv.

Bringéus, N.-A. 2000. "The Red Gold of the Forest." In *Food from Nature: Attitudes, Strategies and Culinary Practices. Proceedings of the 12th Conference of the International Commission for Ethnological Food Research, Umeå and Frostviken*, Sweden, June 8–14,

1998, edited by P. Lysaght, 225–239. Uppsala: The Royal Gustavus Adolphus Academy for Swedish Folk Culture.

Burton, A. 2006. *Archive Stories: Facts, Fictions, and the Writing of History*. Durham, NC: Duke University Press.

Caldwell, M. 2004. *Not by Bread Alone: Social Support in the New Russia*. Berkeley: University of California Press.

Dobson, M., and B. Ziemann, eds. 2009. *Reading Primary Sources: The Interpretation of Texts from Nineteenth- and Twentieth-Century History*. London: Routledge.

Eelmaa, E. 1985. *Toiduainete Külmutamine*. Tallinn: Valgus.

Eestikeelsete Taimenimede Andmebaas. 2014. *Eestikeelsete Taimenimede Andmebaas*. Accessed February 18, 2014. http://www.ut.ee/taimenimed/

Etverk, I., and H. Sein. 1995. "Metsad ja Nende Majandamine." In *Eesti Loodus*, edited by A. Raukas, 402–417. Tallinn: Valgus, Eesti Entsükolpeediakirjastus.

Goldberg, C. 1984. "The Historic-Geographic Method: Past and Future." *Journal of Folklore Research* 21 (1): 1–18.

Hagström, C., and L. Marander-Eklund, eds. 2005. *Frågelistan Som Källa Och Metod*. Lund: Studentlitteratur.

Hietala, M. 2003. "Food and Natural Produce from Finland's Forests during the Twentieth Century." In *The Landscape of Food. The Food Relationship of Town and Country in Modern Times*, edited by M. Hietala and T. Vahtikari, 185–197. Helsinki: Finnish Academy of Science & Letters. Vol 4 of *Sudia Fennica, Historica*.

Jõesalu, K. 2003. "What People Tell about Their Working Life in the ESSR, and How Do They Do It? Source-centered Study of a Civil Servant's Career Biography." *Pro Ethnologia* 16: 61–88.

Jürgenson, A. 2005. *Seened Kultuuriloos*. Tartu: Ilmamaa.

Kalle, R., and R. Sõukand. 2012. "Historical Ethnobotanical Review of Wild Edible Plants of Estonia (1770s–1960s)." *Acta Societatis Botanicorum Poloniae* 81 (4): 271–281. doi:10.5586/asbp.2012.033.

Kasekamp, A. 2010. *A History of the Baltic States*. Basingstoke, NY: Palgrave Macmillan.

King, M. T. 2012. "Working With/In the Archives." In *Research Methods for History*, edited by S. Gunn and L. Faire, 13–29. Edinburgh: Edinburgh University Press.

Klesment, M. 2009. "The Estonian Economy under Soviet Rule: A Historiographic Overview." *Journal of Baltic Studies* 40 (2): 245–264. doi:10.1080/01629770902884284.

Kõresaar, E. 1995. "Questionnaire as a Moulder of Ethnological Source." *Pro Ethnologia* 6: 36–50.

Kusmin, J., and T. Kusmin. 2011. "Metsa Kõrvalkasutus Riigimetsades 20. Sajandil." *Metsa Kõrvalkasutus Eestis, Akadeemilise Metsaseltsi Toimetised* 25: 15–28.

Kuutma, K., and T. Jaago, eds. 2005. *Studies in Estonian Folkloristics and Ethnology*. Tartu: Tartu University Press.

L'Heureux, M.-A. 2010. "Modernizing the Estonian Farmhouse, Redefining the Family, 1880s–1930s." *Journal of Baltic Studies* 41 (4): 473–506. doi:10.1080/01629778.2010.527134.

Lévi-Strauss, C. 1966. "The Culinary Triangle." *The Partisan Review* 33: 586–596.

Lindhagen, A., and L. Hörnsten. 2000. "Forest Recreation in 1977 and 1997 in Sweden: Changes in Public Preferences and Behaviour." *Forestry* 73 (2): 143–153. doi:10.1093/forestry/73.2.143.

Lindquist, Y. 2009. *Mat, Måltid Minne. Hundra år Av Finlandssvensk Matkultur*. Helsingfors: Svenska Litteratursällskapet i Finland.

Linnus, F. 1937. *Rahvateaduslikud Küsimuskavad X. Söögid, Joogid, Maitseained*. Tartu: Sihtasutus Eesti Rahva Muuseumi Kirjastus.

Łuczaj, L. 2008. "Archival Data on Wild Food Plants Used in Poland in 1948." *Journal of Ethnobiology and Ethnomedicine* 4: 4. doi:10.1186/1746-4269-4-4.

Łuczaj, L., A. Pieroni, J. Tardío, M. Pardo-de-Santayana, R. Sõukand, I. Svanberg, and R. Kalle. 2012. "Wild Food Plant Use in 21st Century Europe: The Disappearance of Old Traditions and the Search for New Cuisines Involving Wild Edibles." *Acta Societatis Botanicorum Poloniae* 81 (4): 359–370. doi:10.5586/asbp.2012.031.

Luts, A. 2008. "Loodusvarud Majandamises." In *Eesti Rahvakultuur*, edited by A. Viires and E. Vunder, 107–135. Tallinn: Eesti Entsüklopeediakirjastus.

Lysaght, P. 2000 "Introduction." In *Food from Nature: Attitudes, Strategies and Culinary Practices. Proceedings of the 12th Conference of the International Commission for Ethnological Food Research, Umeå and Frostviken*, Sweden, June 8–14, 1998, edited by P. Lysaght, 11–18. Uppsala: The Royal Gustavus Adolphus Academy for Swedish Folk Culture.

Manoff, M. 2004. "Theories of the Archive from Across the Disciplines." *Portal: Libraries and the Academy* 4 (1): 9–25. doi:10.1353/pla.2004.0015.

Meikar, T., and I. Etverk. 2000. "Metsaomand Eestis." *Metsanduslikud Uurimused* 32: 8–18.

Moora, A. 1981. "Marjad Rahvatoidus." *Eesti Loodus* 6: 567–570.

Moora, A. 2007. *Eesti Talurahva Vanem Toit*. 2nd ed. Tartu: Ilmamaa.

Moore, F. P. L. 2010. "Tales from the Archive: Methodological and Ethnical Issues in Historical Geography Research." *Area* 42 (3): 262–270.

Notaker, H. 2003. "Nordic Countries." In *Encyclopedia of Food Culture*, Vol. 2, edited by S. H. Katz, 558–567. Farmington Hills, MI: Thomson Gale.

Ogborn, M. 2010. "Finding Historical Sources." In *Key Methods in Geography*, edited by N. J. Clifford, S. French, and G. Valentine, 89–102. 2nd ed. London: Sage.

Olsson, P. 2011. *Women in Distress: Self-Understanding Among Twentieth-Century Finnish Rural Women. European Studies in Culture and Policy*. Wien. Vol. 11. Berlin: LIT.

Paal, T. 2011. "Metsamarjade ja Seente Varumine." *Akadeemilise Metsaseltsi Toimetised* 25: 67–72.

Pärdi, H. 1983. *Küsimusleht Nr 168. Loodusandide Korjamine*. Tartu: Eesti Etnograafiamuuseum.

Pärdi, H. 2008. "Talumajandus." In *Eesti Rahvakultuur*, edited by A. Viires and E. Vunder, 75–106. Tallinn: Eesti Entsüklopeediakirjastus.

Pärdi, H. 2012. *Eesti Talumaja Lugu. Ehituskunst ja Elu 1840–1940*. Tallinn: Tänapäev.

The Plant List. A Working List of All Plant Species. Accessed February 18, 2014. http://www.theplantlist.org

Piiri, R. 2002. *Küsimusleht nr 214. Toidukultuur Nõukogude Ajal*. Tartu: Eesti Rahva Muuseum.

Piiri, R. 2006. "See Varumise Harjumus–Toidukultuurist Nõukogude Eestis." *Eesti Rahva Muuseumi Aastaraamat* 49: 49–90.

Pouta, E., T. Sievänen, and M. Neuvonen. 2006. "Recreational Wild Berry Picking in Finland—Reflection of a Rural Lifestyle." *Society & Natural Resources* 19 (4): 285–304. doi:10.1080/08941920500519156.

Raig, I. 1981. *Teie Isiklik Abimajapidamine. Sotsiaal-Majanduslik Aspekt* [Your Personal Supplementery Economy. A Socio-economic Aspect]. Tallinn: Valgus.
Raun, T. U. 2001. *Estonia and the Estonians*. Stanford, CA: Hoover Institution Press.
Riessmann, C. 2008. *Narrative Methods for the Human Sciences*. Los Angeles, CA: Sage.
Shephard, S. 2006. *Pickled, Potted, and Canned: How the Art and Science of Food Preserving Changed the World*. New York: Simon & Schuster.
Sion, V. 1947. *Küsimusleht nr 43. Seene-, Marja-, Pähkli-ja Muu Taimtoidu Korjamisest*. Tartu: Eesti Rahva Muuseum.
Skott, F. 2008. *Folkets Minnen. Traditsionsinsamling i Idé Och Praktik 1919–1964*. Göteborg: Göteborgs Universitet.
Spuhl-Rotalia, J. G. 1898. *Kodumaa Marjad: Täielik Õpetus Kodumaal Kaswawaid Marjasid ja Puuwiljasid Kaswatada, Sisse Korjata, Müügile Wiia ja Alal Hoida, Neist Weini, Likööri, Limonaadi, Moosi, Sahwti, Marjasülti, Siirupit, Marmelaadi, Kompoti, Salatid, Suppi, Kookisi, Teed, Kohwi, Äädikat jne. Walmistada, Nende Terwisekasulisi ja Terwisekahjulisi Omadusi Tunda ning Neid Arstliselt Tarwitada*. Viljandi: A. Peet.
Steedman, C. 2001. *Dust*. Manchester: Manchester University Press.
Svanberg, I. 2012. "The Use of Wild Plants as Food in Pre-industrial Sweden." *Acta Societatis Botanicorum Poloniae* 81: 317–327. doi:10.5586/asbp.2012.039.
Tael, T. 2006. *Mälu paberil: Eesti Rahva Muuseumi Korrespondentide Võrgu 75. Aastapäevaks*. Tartu: Eesti Rahva Muuseum.
Teplyakov, V. K. 1998. *A History of Russian Forestry and Its Leaders*. Collingdale, PA: Diane.
Troska, G., and A. Viires. 2008. "Söögid ja Joogid." In *Eesti Rahvakultuur*, edited by A. Viires and E. Vunder, 264–278. Tallinn: Eesti Entsüklopeediakirjastus.
III. Põllumajandusloendus. 1939. *A. 1. Vihik = Données Du Recensement Agricole De 1939* [1940]. Vol. 1. Tallinn: Riigi Statistika Keskbüroo.
Värv, E. 1990. "Correspondents' Contribution to the Estonian National Museum." *Cahiers Du Monde Russe Et Soviétique* 31 (2): 287–293. doi:10.3406/cmr.1990.2227.

"IS THAT HUNGER HAUNTING THE STOVE?"

THEMATIZATION OF FOOD IN THE DEPORTATION NARRATIVES OF BALTIC WOMEN

Leena Kurvet-Käosaar

This article provides an analysis of the ways in which food has been thematized in the deportation narratives of Baltic women (primarily narratives of the 1941 mass deportations). Based on a posthumously published diary of an Estonian woman who died in her deportation location in 1945, the well-known deportation memoir of Lithuanian woman Dalia Grinkevičiūtė as well as a number of deportation stories of Baltic women collected and published in the early 1990s, the article focuses in particular on the following topics: food and identity, communal networks of care, different means of procuring food, and hunger.

Personal accounts of Stalinist repressions in the Baltic states, including deportation and labor camp narratives, have emerged as important sites for constructing collective memories that underscore the violent nature of the regime and testify to the capacity of the Baltic people to survive in the face of the destruction of national communities both on sociocultural and material level (Kirss 1999, 23–31, 2004, 13–18; Hinrikus 2004, 62–77; Kurvet-Käosaar 2005, 59–78; Lazda 2005, 1–12; Budrytė 2010, 331–50). Whether in the form of written narratives or oral history interviews, these accounts became the cornerstone of various life history archives in the Baltic states that today house extensive corpora of life narratives. The overall number of women's life stories equals or even exceeds those of men as women and children were more often deported to those areas of the Soviet Union where living conditions, although varying in harshness, made survival more likely than in labor camps. Therefore, more women survived to record their experience in the late 1980s and early 1990s when several large-scale life story and oral history projects encouraging writing and telling about the experience of the Soviet regime were launched in the Baltic states. Women's narratives contribute significantly to the more general gaining of a voice of women in the Baltic discourses of memory and history and underscore the relevance of the practices of everyday life to the preservation of national culture (Kurvet-Käosaar 2012, 91; Kirss 1999, 24).

Survival in Siberia depended crucially on the availability of food. In an extensive number of life stories, particularly those by women, food shortage and hunger constitute an important thematic thread tying together the deeply traumatic experiences of exile and forced labor. Drawing on various published deportation and Gulag narratives of Estonian, Latvian, and Lithuanian women, my article seeks to analyze the thematization of food, focusing on the role that descriptions of procuring, preparing, and consuming food, as well as the practical and social means of coping with food shortage and hunger in the representations of deportation. In addition to the emphasis on bodily nutrition, food in the Baltic women's narratives is often constructed as metaphors of survival and perseverance. In her consideration of the deportation narratives of Polish women, Katherine Jolluck argues that, in addition to "physically debilitating conditions and treatment," deportation constituted a radical threat, "a sense of self and belonging to a community" to both men and women. Yet it can be viewed as a "compounded injury" for women who traditionally have been defined predominantly through the domestic sphere. Therefore the process of "striving to recreate a home" can be viewed as a key element in women's adoption to and preservation of themselves in a harsh and hostile environment (Jolluck 2002, xx). In the life stories of Baltic women, food becomes a particularly significant aspect of (re) creating home, visible on the most basic level via efforts to feed one's family, through attempts to preserve the rituals of domesticity and family life and different networks of care and support that often developed around procuring and preparing food.[1]

As the life stories make visible, hunger "haunts the stove" (Nagel 2007, 68) of more or less everyone deported in 1941 or shortly after, thus, posing one of the most common risks to one's life and well-being. Women's deportation stories make visible in great detail the ways in which the deportees dealt with food shortages, as well as the ways in which food is thematized in their life writings. The thematic foci concerning food range from making adjustments in culinary practices and displaying considerable resourcefulness in procuring food products, to suffering from malnutrition and hunger for extended periods of time and witnessing deaths caused by hunger among family members and fellow deportees. In many narratives, it is particularly these descriptions of suffering from and succumbing to hunger that the trauma of deportation emerges most clearly. However, unlike many other aspects of the deportee situation that were beyond one's control, hunger could be fought in many different ways, such as, by re-distributing the scant daily rations, procuring extra products through bartering personal belongings, or performing various services for the local people (e.g., knitting, drawing), through the use of all natural resources available, and creating new recipes that made the most efficient use of the available products. As a central aspect of everyday life that could and had to be tended to on a daily basis, food in many narratives becomes an important thematic instrument, a strategy for making visible the process of "wresting personal victories, however small, from history"

(Skultans 1997, 19), and positing the experience of deportation "as another in the sequence of life's many hardships, to be borne with resourcefulness, skill, as honestly and humanely as possible" (Kirss 2005, 20). In my article, I will focus on the thematization of food via different topic areas that have emerged most clearly during my research. First and foremost, the life narratives demonstrate the importance of food in terms of identity-forming processes. Discussion of culinary habits and traditions back home and descriptions of authors' adjustment of them in deportation areas emerges in the life stories not only to highlight strategies of pure physical survival but just as importantly as a way of making life more manageable and even enjoyable and to build up and maintain a sense of community. Second, different means of procuring food that constitute a prominent thematic focus in the life stories make visible the importance of various semi-official and informal networks of care both along the lines of national identity and within the local community that are crucial for people's survival in Siberia. Relating to communal networks of care is the question of bartering – exchanging items of clothing and valuables for food with the local people. Here, the emphasis is both on bargaining skills that the deportees developed as an efficient means of survival as well as on the unequal power dynamics between them and the local people who were in the position to set the terms of the exchanges. Another aspect relating to food thematized in the narratives is "pinching" food (most frequently collecting leftover, often hardly edible crops and grains from the fields). Via this ethically highly charged aspect the authors strive to make visible the graveness of their condition as well as the extreme injustice to which they were subjected to in Siberia. Finally, I will discuss the most common and most prominent theme relating to food in the narratives – that of food shortage and hunger – that in turn, importantly relates to the ways in which such "culture of hunger" emerges in the narratives as a testing ground for survival, fight against erosion of humanity, and preservation of a livable and meaningful every day in the face of the extreme.

The article was inspired by the posthumously published diary of an Estonian woman Erna Nagel, "*Olen kui päike ja tuul Erna Nageli päevik*" [The *Diary of Erna Nagel. I Am Like the Sun and the Wind*] (2007). Deported with her mother in June 1941, Erna Nagel ended up in the village of Bondarka near Kargassok in the Siberian taiga and died in her deportation location in January 1945. In general, very few records of the immediate everyday experience of deportation have been preserved. First and foremost, this diary's preeminent thematic focus on food makes it a unique and valuable source for the current research. The extensive and detailed record of the deportation experience of the Lithuanian woman Dalia Grinkevičiūtė, written in 1949–1950, first published in 1997 and in 2002 translated as *A Stolen Youth, a Stolen Homeland*, has provided another important nexus for the article (Grinkevičiūtė 2002). While Erna Nagel's diary offers extensive insights into the ways of coping with food shortage and hunger, Dalia Grinkevičiūtė provides a detailed record of most merciless starvation to which the Lithuanian deportees were subjected on the uninhabited island of Trofimovsk in the Laptev Sea. For further elaboration of the main subthemes of the article, I have used the life story collections *14. Juuni 1941. Mälestusi Ja Dokumente* [14 June 1941. *Memories and Documents*] (Laar 1990), *Eesti Rahva Elulood II* [*Life Stories of the Estonian People, Volume II*] (2000) and *Me Tulime Tagasi* [*We Came Back*] (1999) for

Estonian women's life stories, and *We Sang Through Tears: Stories of Survival in Siberia* (1999) for Latvian women's life stories.

The deportees were sent to the most underdeveloped areas and zones of harsh climatic conditions. This had particularly serious effects on the victims of the first major wave of deportations of June 1941, among whom the death rates are estimated to have been around 60% (Mertelsman and Rahi-Tamm 2009, 310). At the beginning of the journey to Siberia, men were separated from their families and sent to forced labor camps, where a majority of them perished or were executed. Estonian women, children, and the elderly were sent to the Kirov and Novosibirsk oblast (Rahi-Tamm 2007, 6), Latvians to the Krasnoyarsk and Tomsk districts (Geka 2012, 7), and Lithuanians to Altai, Novosibirsk, Komi ASSR, and the region of Gurjec in Kazakhstan (Anušauskas 2013). In 1942, over 2000 Lithuanian deportees from the Altai region were taken to the north of Jakutia, to the islands of the mouth of the river Lena in the permafrost region (Anušauskas 2013). The 1941 deportees were also hit very hard by the crop failure of 1941–1942 that was followed by a severe famine (Rahi-Tamm 2007, 50). Since the climatic regions (taiga, tundra, and areas beyond the Arctic Circle) where the deportees were taken differed greatly in terms of what could be cultivated or gathered from nature in the area, the narratives do not allow for extensive generalizations in terms of the availability of food in a certain climactic zone. In principle, those who were exiled to harsher climatic conditions suffered more severely from malnutrition and hunger, but the situation depended to a great extent on the agricultural management skills of the kolkhoz where the deportees were assigned, as well as the attitude of the collective farm toward the deportees. The question of the relationship between climatic zones, land cultivation developments, and the availability of food in the areas of deportation constitute a large-scale research project of its own, which is beyond the scope of the present article, where the focus is on the role of the thematization of food in deportation and labor camp narratives.

The deportees were sustained on arbitrarily distributed scant food rations dependent on unrealistic work quotas. Listing "the regulation of prisoners' food" among the "camp administration's most important tool[s] of control," Anne Applebaum regards the distribution of food in the Gulag as an "an elaborate science," calculated on the basis of "the minimum quantity of food necessary for prisoners to continue working," but often appended according to the availability of food and various other considerations (Applebaum 2003, 207–08). Although the situation of the deportees differed in several aspects from that of camp inmates, such logic of food distribution also applied to them. Both for the camp inmates and the deportees, food rations depended on fulfilling work quotas, with calculations about the amount of work done and rations allotted depending on the superiors' goodwill and conscientiousness. Those members of the deportee community (children, the elderly, and the sick) who were unable to work sometimes received reduced rations, but often were completely dependent on the working family member(s) (Rahi-Tamm 2007, 44). In comparison to camp inmates, deportees had somewhat greater mobility, so they could, for example, exchange their belongings for food in the more well off villages and procure extra food (edible plants, berries, mushrooms) from the forest.

Food and Identity

In the life stories, the experience of deportation is described in terms of a sudden and violent break with all familiar contexts of everyday life, and further absence or deficiency of these contexts to a degree that can often be described as inhuman. Yet even under such extreme circumstances, food emerges as "a decisive element of human identity and as one of the most effective ways of expressing that identity" (Montanari 2006, xi). Cara de Silva, editor of the cookbook compiled by the women of the concentration and transit camp of Theresienstadt (Terezín) advances a similar view, maintaining that "[o]ur personal gastronomic traditions … are critical components of our identities. To recall them in desperate circumstances is to reinforce a sense of self and to assist us in our struggle to preserve it" (de Silva 1996, xxvi). In the deportation narratives, meals back at home are sometimes mentioned to underline the stark contrast with their current life and express a longing for the life left behind. This is clearly visible in the diary of Erna Nagel, who, on several occasions, juxtaposes the meals she used to have back home and her current food options as ways to express her sense of loss and trauma:

> How painful it was to recall Ülenurme this morning (…) I imagined so vividly all the traditional Sunday dishes that it almost seemed to me that I can smell and taste them. I saw my aunt carrying a big load of curd pastries to cool in the living room and I didn't forget scrambled eggs and ham with horseradish [we used to have] on Saturdays. Thoughts fit for a slave of one's stomach! (Nagel 2007, 102)

Highlighting both the dishes that were eaten on special occasions and prepared according to family traditions, as well as the peaceful and festive atmosphere of the meals that brought together the whole family, Erna is not only concerned with food shortage that she suffered at that time but also the loss of home and the familiar, taken-for-granted contexts of her life. The fact that the loss of home is articulated through a reference to food underlines the importance of cooking and eating as essential elements of one's well-being.[2]

Descriptions of dreaming of food, discussing different ways of how certain dishes were prepared back home and exchanging recipes with compatriots can also be found in other Baltic women's deportation narratives. These occasions are presented as ways to strengthen one's belonging to a community and similarly to the women of Terezín, that is, "bring comfort amid chaos and brutality" (de Silva 1996, xxvi). Lidija Vilnis, a young Latvian woman who was pregnant at the time of deportation and who suffered from extended periods of hunger in her deportation location of Yeryomina on the Yenisey River, mentions alleviating the torment of hunger with the help of sharing recollections of familiar culinary traditions with other Latvian women:

> As soon as a few of us women were together, we would fantasize, "cook" and "bake" whatever we remembered or made up. We would all listen enthusiastically and write down these recipes … this is how all the photographs I had with me ended up with recipes in the back. (Vilnis 1999, 98)

This example conveys confirmation of double communal identification: a sense of national unity is created via a specific sphere of life and skills characteristic in such capacity only of women. Similarly to the cookbook compiled by the women of Terezín, the sharing of recipes testifies to "the women's desire to preserve something of their past world, even as that world was being assaulted, and it attests to their own recognition of the value of what they had to offer as women – the knowledge of food preparation" (Hirsch and Spitzer 2006, 356).

Together with the descriptions of daily efforts of procuring, preparing, and consuming food, these descriptions can be viewed as proof of what Tzvetan Todorov in his consideration of moral life in the concentration camps has referred to as "acts of ordinary virtue" (Todorov 1996, 59). In his discussion of the range of human response to the Nazi *lagers* and the Soviet *gulag*, Todorov affirms the "continuity between ordinary experience and that of the camps ... and thus the pertinence of the same moral questions to both worlds" (1996, 40). According to Todorov, despite the extreme circumstances of the camps, moral life as an essential component of humanity does not cease there. The essential proof of its existence can be found in the acts of ordinary virtue that, in turn, attest to the continuation and importance of at least some form of "normalcy." In Baltic women's deportation and labor camp narratives, extended networks, practices, attachments, and identifications relating to food make visible its central importance in creating a habitable everyday life. Such manner of conveying the deportation experience testifies to the survival of the ordeals of the repressions for the reader and also functions as an effective emotional coping mechanism for the authors themselves. Recreating the daily rhythm of their lives through a focus on everyday practices can provide reassurance for the authors about the manageability of the harsh and potentially traumatic experience, as well as facilitate its retrospective narration many decades after its occurrence.

The relevance of food in the process of creating habitable, everyday life is visible in a particularly clear manner in the diary of Erna Nagel. Dominated by descriptions of procuring food, and meals prepared with the products at hand using both makeshift Estonian as well as local recipes, the diary can be viewed as an outstanding example of coping with deportation by adapting to the new circumstances with relative success. The diary also testifies to acquiring the skills, most importantly those of procuring and preparing food necessary not only for survival but even for a degree of enjoyment. At the same time, the ironic tone of the diary also makes visible the author's constant awareness of the extent to which her present life differs from her past one, and the triviality and absurdity of the concerns of her daily life with its constant focus on food. Yet the repetitive entries that also constitute a record of her daily struggle for survival with the ironic tone faltering with her gradual regression to hunger can be interpreted as textual evidence of both suffering from and coping with the trauma of deportation.

"This is how my days pass by here, I eat and think about what to eat. When my stomach is full, I think of what to eat tomorrow and then I want to sleep – this way I need to think less," Erna notes on 14 July 1941, exactly one month after being deported (Nagel 2007, 9). Erna's ironic remark about concerns centering on food taking over her life does not aim at highlighting this theme *per se* but rather draws attention to the abrupt and limiting change in her life that used to be filled with issues concerning her studies, active social life, and romantic interests. Although Erna expresses hope of being able to return home soon, the first sections of her (preserved) diary indicate that she also perceives that the change in her life is both substantial and irreversible. Making a record of the meals she and her mother have each day can be viewed here as a textual means of handling the overwhelming and unmanageable questions of her future life through an ironic focus on the details of her daily routine. As her situation gradually worsens, detailed descriptions of meals become an important means of fighting hunger. Despite the gradual improvement of their situation by the end of 1943, Erna's mother dies in July 1944. Though food, and in particular festive meals prepared for various communal celebrations, provides some consolation to Erna in her mourning, she never fully recovers from her mother's death. Erna dies in January 1945 after being caught under a falling tree during timber cutting work.

Communal Networks of Care

In many women's deportation narratives, different means of procuring food constitute a prominent thematic focus, making visible the importance of various semi-official and informal networks that are crucial for people's survival in Siberia. Often the capacity of the deportee community of one nationality to stick together and distribute food resources among themselves is viewed as essential strategy for survival that also provides strong moral support. A typical example can be found in the narrative of an Estonian woman, Evi Tallo, who came from a relatively wealthy background in Rakvere, was deported twice and sent to Kargasok (taiga). "One early spring we were in quite a hopeless situation," Evi writes. "Mother's friends from Rakvere learned about this, came with a sledge and brought us potatoes, flour, and had even collected some money. We were safe again!" (Tallo 1999, 150).

Similarly, family provides an important safety net. Lidia Vilnis, a young Latvian woman who was expecting a baby when she was deported, attributes being able to stay alive to her husband despite "the famine [that] was appalling. I still had a small, nickel-plated watch, which I offered [the doctor's assistant]. I 'sold' it for a loaf of bread, a liter of oil and a kilo of dark noodles. That was how, at that critical time, my husband's gift saved my life" (Vilnis 1999, 100). Similar to the other families deported in 1941, Lidia was separated from her husband during the journey to Siberia and never saw him again. Her decision to emphasize familial support when recounting the episode almost half a century later serves the purpose of highlighting the endurance of communal and familial ties that in her narrative form a stark contrast with the hostile contexts of her everyday life.

While Erna Nagel also mentions the support she receives from her fellow Estonian deportees, in particular her best friend Hilda Põltsamaa, she also records conflicts and rivalry between her compatriots that she attributes to limited circumstances that forced people to protect their interests by all available means: "What a life, when even our own people turn into wolves" (Nagel 2007, 132). This is how Erna concludes an account of tense relations between two brigades of Estonian forestry workers in February 1943; she continues to list occasions when fellow Estonians have kept to themselves information about food products that were distributed in the village shop so that they could get bigger rations at the expense of the others. Erna Nagel's diary also contains numerous mentions of receiving help from local people of different ethnicity and background, as well as sharing her food with the others whenever possible. On her way to the bazaar in a near-by village, Erna often visits an elderly couple, afterwards reflecting on their unfailing kindness and generosity in her diary:

> As if sensing my hunger … these kind people treat me to a hardy meal. [I get] a sizable portion of porridge, a bowl of milk and another bowl of fresh cabbage. I eat with a good appetite. … Noticing my appetite, the landlady, with her heart of gold, pours more milk and [I eat] until my stomach is completely full. My mood also rises 100 percent. (Nagel 2007, 218)

An Estonian woman, Mirjam Kaber, who was in 1941 as a very young child deported with her mother to Molotovsk in the north of the Arkhangelsk Oblast (taiga), also recounts an instance of surviving thanks to the help of their local landlady: "I cannot but remember with utmost respect our landlady Anna Sergejevna, if it wasn't for [her], my mother indeed would have stayed forever in the sandy shores of the Voja River" (Kaber 2000, 152). Women's deportation stories provide ample evidence of the ways in which food assumes a significant role as the facilitator of processes of socialization that include people of their own nationality, fellow Balts, as well as local people – and sometimes even Soviet authorities. Outlining different networks of care in the narratives serves the purpose of highlighting national unity as well as the capacity for resourcefulness necessary for survival. In addition, it also offers proof of connectivity and a sense of community as a "deeply held value for women" in general that is related to "female concepts of adulthood and virtue" (Long 1999, 49, see also; Kurvet-Käosaar 2005, 59–78).

Bartering

Even more frequent than examples of being helped by the local people are descriptions of exchanging items of clothing and other valuables for food with local people, who dictate their own terms for these exchanges. "The day passed under the star of bartering," notes Erna on 22 March 1942, "Mother sold our mistress 2.5 meters of satin and striped satin (material for our dear father's suit) for 4 buckets of potatoes, and her light coat for 250 rubles and 12 buckets of potatoes" (Nagel 2007, 75). Such bartering took up a great deal of energy and time for both her and her mother. In addition to selling their things in their own village, they frequented the bazaar in a nearby town, went door to door in other villages, and distributed information about items they had for sale among people.

Often bartering did not proceed as smoothly as in the cited example. For example, once when Erna went to collect some eggs as payment for her drawings, the owner of the household asks 15 rubles for 1 egg, apparently 15 times the price of eggs, and then refuses to sell them even at that monstrous price. "I quickly say good-bye and leave, as my eyes start brimming with tears. Poor me, I never have gotten used to begging" (Nagel 2007, 80). As the description of the incident makes clear, the owner of the farm did not raise the price of the egg in order for Erna to really pay the unrealistic price, but rather to make her beg for a better deal. However, as Erna realizes, this would not have really secured the deal but only provided the owner the possibility to humiliate her and to reconfirm the power hierarchies among the local people and the deportees.

In general, the possibility of exchanging personal belongings for food provided a valuable means of avoiding hunger and allowed the authors to demonstrate their capacity to develop (bargaining) skills necessary for survival. Although descriptions of bargaining often highlight power hierarchies between the deportees and the local people, examples of fair and even generous deals can also be found from the narratives. Such exchanges make visible the empathic and supportive attitudes of the local people who seem to be motivated first and foremost by a desire to improve the miserable condition of the deportees. Herta Kaļiņina, who came from a rural background near Liepāja and was deported at the age of 15 to a location near Igarka (taiga) with her mother and three siblings, recalls being left completely without rations by the kolkhoz. Her experience of bartering in the nearby, more prosperous Tatar kolkhozes is entirely positive: "the Tatars always fed us and even gave us crusts of bread to take with us" (Kaļiņina 1999, 74). Dalia Grinkevičiūtė's life story contains the most desperate descriptions of horrible suffering from hunger on the Island of Trofimovsk, yet even in the most extreme circumstances, some help was available. The Evenks from the neighboring islands, who survived on fishing and hunting Polar foxes, were prepared to help the deportees, but were stopped by the Soviet authorities: "They could help and they wanted to help. They had provisions of fish and enough dog sleds to bring us to their heated yurts. But our supervisors would not allow it and in this way condemned us to death" (Grinkevičiūtė 2002, 45).

"Pinching" Food

"Went picking berries with Miss Põltsamaa. Sneaked across the field and into the woods like thieves to find some cloudberries in the swamps," Erna Nagel writes on 18 July 1942 (Nagel 2007, 104). This entry shows that in Erna's relocation village, berries are also considered the property of the state and picking them a theft, although Erna refuses to accept this. On another occasion when she returns from harvesting potatoes with "brimming pockets," this triggers in her ethical contemplation on the topic of stealing: "What has become of us? We no longer have respect for the property of the others. What was the commandment, I think, thou shall not steal. But the potatoes are so tempting" (2007, 109).

Many narratives describe in detail both the moral implications and dangers involved in procuring extra food from the kolkhoz fields, during transport or from nature. Sometimes the narrators also employ a particular vocabulary for referring to such activities. For example, in Estonian life stories, the words *kiputama* and *krattima*, both humorous equivalents of the English word "pinch" are used, demonstrating partial lenience toward such means of procuring extra food. Palmi Sink, an Estonian woman, deported at the age of 14 together with her sister, comments on the word *kiputama* in the following manner: "A nice word, better anyway than the word 'to steal.' How could one talk about theft when people are forced to toil from morning till night almost without any pay" (Sink 1999, 47). When Lidia Vilnis observes how "corn was scattered all along the roads" during transport and "cattle were free to … eat from the piles of wheat," the idea of theft does not even cross her mind when she "calmly put[s] some [corn] in her pocket," in full view of everyone. As she is "nearly put to prison"

(Vilnis 1999, 93) for that pocketful of corn, she soon learns the rules the hard way but considers them immoral with respect to the most reckless neglect and deprivation the deportees have been made to suffer from. Other narratives also emphasize that resorting to such means is triggered by extreme food shortage, and what is taken is mostly harvest leftovers and products spilled during transport.

Quite frequently, the fact that food was procured via such means in order to feed family members or fellow deportees is outlined in the narratives as a mitigating circumstance. Valija Kalēja, a teacher who was arrested for lending an English grammar book to a student, was sent to a forced-labor and re-education camp Balagannaya by the Sea of Okhotsk near Magadan in the permafrost region (Kalēja 1999, 33). She recounts a period when she was working as a pig-herder some distance away from the camp. Although the work was quite hard, she was not suffering from hunger but she was concerned about her compatriots who were inmates of the same camp, and so decided to help them by bringing them some extra food. She describes in detail making a kind of pâté of the liver of Siberian salmon, which was abundant at the village near the delta of the Tauya River: "Taking things into the camp was risky because the guards could search you. I sewed two small, narrow bags, which I would tie around my waist under the clothes. One contained soybean cakes, the other some liver pâté" (1999, 38). Valeja was never caught during the summer. In the fall she was assigned to take care of the sows with young piglets at night (1999, 39). She could get some extra food from the kitchen where pig feed was prepared and where the "food was more varied and filling than the one in [the] canteen" (1999, 40). Valeja continued sneaking "an extra morsel" for her compatriots until it finally cost her the job. "Well, what do you expect?" she asks rhetorically and without the least regret justifies her actions in moral terms: "A morsel of food had been taken from a satiated pig and given to a hungry human being" (1999, 42).

If Valeja describes taking extra food to her fellow deportees as a risk she was willing to take without much fear of punishment, numerous other authors are rather concerned with the risks involved in such activities. In her narrative, Palmi Sink describes measures taken against getting caught as well as the risks involved:

> When we lived in Bašina, sometimes we would go "pinching" potatoes in the evening. The darker and stormier the night, the better as then the guards can't hear you. (…) We knew that … one could be put to prison but empty stomach didn't care much for that.… This kind of "pinching" was very risky. There were stories about a woman who had taken four beets … and had been sentenced to four years of prison for that. She died in prison. (Sink 1999, 51)

A Culture of Hunger

In the deportation stories, food shortage and hunger constitute the main forces shaping culinary practices during deportation and determining the ways in which food is handled in the narratives. Although not every single deportation and labor camp narrative by authors who were deported in 1941 or shortly after recounts periods of suffering from hunger and starvation, these conditions are raised in some form in the majority of them, often with reference to the death or near-death from a combination of hunger and exhaustion from work of fellow deportees and family members. Lilija Bīviņa, who was sentenced to a labor camp in Severnaya Dvina, describes how one night she barely made it back to the barracks through a snowstorm. Exhausted, without hope and about to give up on life, she suddenly noticed a piece of bread that had been left on her pillow:

> But then you notice something on the pillow. Your eyes can no longer see, but they've given a secret message to your brain, but it no longer wants to carry out its function.... There is bread on the pillow! Bread? No, rather the helping hand and a loving heart of a friend.... Have you ever held such a piece of bread in your hand? And could you simply eat it? ... I received one like that. I didn't eat it, but pressed it to my heart. (Bīviņa 1999, 118)

Lilija's life story is rare among deportation and labor camp narratives for the absence of the relief and consolation of a survival story, highlighting instead its unbearable nature that testifies to trauma's "endless impact on life" (Caruth 1996, 7; Kurvet-Käosaar 2013, 138–41). The episode where Lilija describes being the recipient of "an act of caring" (Todorov 1996, 72) is among the very few occasions in her narrative that expresses faith in the preservation of humaneness, even if it seemed to be eroded. Although Todorov points out that instances of sharing food were the most common acts of caring in the camps, the episode recounted here carries extra emotional weight for the author. The added exhaustion of getting caught in the blizzard, numbing her senses, and paralyzing her mental capacity has rendered her almost completely apathetic to any kind of communication. In a situation, where "words would have failed," the piece of bread that Lilija "pressed against [her] heart" sparks in her a ray of hope "through the dark nothingness" (Bīviņa 1999, 118).

Lilija's narrative as well as several other Baltic deportation and labor camp stories confirm Anne Applebaum's claim about the sacred status of bread in the context of hunger and starvation, along with "a special etiquette [growing] up around its consumption" (Applebaum 2003, 213). In Lilija's narrative, bread restores the author's faith in humanity and provides her with moral strength to survive. Similarly, Estonian woman Palmi Sink, who had to manage alone with her teenage sister in the Tomsk district (taiga), highlights the symbolic significance of bread for an old man dying of hunger: "I remember how the old Uustalu died .... He held a piece of bread in his hand. This was the most valuable thing during these harsh times" (Sink 1999, 45). In her narrative that recounts the death through hunger of hundreds of fellow deportees, Lithuanian woman Dalia Grinkevičiūtė describes a similar situation with the exception that the young boy does not survive to receive his last piece of bread but "die[s] waiting with his hand stretched out" with the "bread [put] into his hand, when he was already dead" (Grinkevičiūtė 2002, 54). In the last two examples, survival by procuring enough food for the members of the immediate deportee community could not be ensured, yet the symbolic significance of bread is employed to highlight the survival of human dignity. In the narratives, these acts function as a means of empowerment and a way of coping with the trauma of the deportation experience.

In many narratives, hunger becomes visible not only through an emphasis on food shortage but also through desperate efforts made to procure anything edible that in its extremity renders previous conceptions of dietary and culinary practices invalid. "Our mothers went to look for edible grass" (Nagel 2007, 153), Erna reports laconically on a Sunday in May 1943. Picking edible plants from nature provided valuable sustenance against hunger: instances of picking and preparing edible plants (such as, for example, nettle and goutweed) can be found in numerous deportation narratives. Several narratives outline that in cases of extreme hunger, every little leaf of grass was eaten, so that the ground looked as if it had been ploughed. As an example, the female author of an anonymous life story, who was deported as a very young child, attributes her survival in a quite matter-of-fact manner to her capacity to eat anything: "Everything that was edible, I consumed. There was not a single blade of grass growing around our house – just bare soil. It looked as if young cattle had visited the site" (*Lapsena Külmal Maal 1990*, 28). Survival in these circumstances comes down to the ability, in any way possible, to ensure physical survival. Within the context of the gloomy and tragic mood of the narrative as a whole, this observation also functions to highlight the most horrible suffering that is the dominating feature of the narrative.

Although the depiction of severe food shortages and hunger commonly highlight the limits of endurance and resourcefulness of the deportees, it is remarkable that even such extreme conditions are sometimes depicted with a sense of satisfaction and even enjoyment. An example can be found in Erna Nagel's diary; this case describes making buns or pancakes out of potato peel: "these '*lepjoškas*' made of potato peel now taste so good" (Nagel 2007, 223). Despite the fact that under normal circumstances, eating potato peel would have been considered inconceivable, it became a useful and nutritious addition to the daily diet, with ingredients varying also according to the greenery, often referred to as "grass" that could be found in nature (Nagel 2007, 109). In a similar vein, Evi Tallo describes secretly picking rotten potatoes on the field that were then prepared into some kind of buns and then baked on top of the stove: "When we also happened to have some flour, it tasted almost like a dessert" (Tallo 1999, 153), she comments. Resorting to extreme means of procuring food is not viewed as degradation or humiliation, but as a positive, even enjoyable measure that also strengthens the sense of solidarity among the deportees and highlights their survival skills. "Soup again. Oh all that water that we are forced to consume," Erna complains in April 1944 and specifies in detail:

> In the morning, three bowls of soup, a few potatoes and the leanest of gruel, or flour or sieved potatoes, sometimes a little bit of fish, a bone with some meat on it or some cabbage … two–three bowls of soup for dinner and two for supper. And so from day to day. Constantly empty stomach makes me quite nervous these days. (Nagel 2007, 223)

On and off, food shortage is a dominating feature of Erna Nagel's diary from almost the beginning of her sentence in 1941 until her death in January 1945. She repeatedly expresses frustration about her constantly empty stomach and becomes increasingly indifferent toward exchanging for food personal belongings reminding her of her past life. Occasionally, she mentions health problems that can be attributed to malnutrition (dizziness, weariness, upset stomach), yet the word hunger is mentioned in the diary only a few times. One of these instances is quite early in the diary, on 27

February 1942, when a foreboding of a dark future suddenly overwhelms Erna, spending an afternoon alone at home:

> We need potatoes; there is not enough bread – and what is 200 grams a day? I fall on my knees and pray. What if my dear mother dies? Something haunts the stove, hunger perhaps? May it be that our real torment is only about to begin? My mother went, two of my dresses wrapped in her pretty scarf. (2007, 68)

The lack of ironic undertones, as well as an incoherent and rambling quality, makes the entry distinctively different from the majority of entries in the diary. It offers something akin to a flash-forward of life: Erna correctly predicts that the haunting shadow of hunger will follow them for years to come. Her diary, however, soon resumes its ironic and at times humorous tone, displaying a marked determination never to be overcome by hunger. If in real life, hunger may be more difficult to fight, Erna can at least keep it off the pages of her diary, counting in a systematic manner every bite she eats and every sip she drinks, demonstrating that not only is she able to conquer hunger but that such life is altogether not devoid of culinary enjoyment.

Descriptions of a constantly empty stomach, which in their intensity push to the background all other concerns, abound in the deportation and labor camp narratives. The examples provided demonstrate the resourcefulness of the deportees and the capacity to cope with the situation both in practical and affective terms. Yet those narratives that depict the most extreme lack and deprivation, or criminal neglect, such as Dalia Grinkevičiūtė's account of her time on the Island of Trofimovsk, also highlight not only terrible suffering but also gradual erosion of humanity. She describes a man who, as a result of hunger, became insane and was overcome by an obsession for food, and so ceasing to divide up his rations over days, prepare, or even cut or thaw it. In her account of him eating a raw frozen fish with "fish intestines … hanging from his mouth … blood streaming over his hands" (Grinkevičiūtė 2002, 53), he no longer resembled a human being but a crazed animal. This description is provided as an extreme example of the effect of hunger that Grinkevičiūtė clearly condemns; yet only a few pages later she describes the situation of her own family using similar imagery. When her mother collapses of hunger, Dalia and her brother realize that like "hungry beasts, frozen and tortured" they had "snatched the bread" their mother had given them "without asking if she had eaten herself" (2002, 61). In Grinkevičiūtė's narrative, this realization functions as a powerful occasion to commemorate the utter dedication and unselfish love of her mother and her willingness to sacrifice herself for the sake of her children: "Mother, all heroes paled in my imagination, unable to stand the comparison with your strength, love and heroism… This is the essence of a mother's heart" (2002, 61). The description of her mother's failing health and her heroism constitutes one of the testimonial cores of the narrative where the author makes visible the utter inhumanity and cruelty of the regime and its officials, as well as the persistence of moral virtues and standards of humanity that is presented as the key to survival.

Grinkevičiūtė's narrative, as well as several other deportation and Gulag narratives of Baltic women, testify to the most extreme condition of hunger that cancels out any resemblance of normal everyday life that the deportees had managed to build up despite the harsh conditions, thus, addressing the question of the erosion of humanity.

In her discussion of "*dokhodyagi*, [those] starving to death" in the Gulag system, Ann Applebaum points out that these people became "the physical fulfillment of the dehumanizing rhetoric used by the state: in their dying days, enemies of the people ceased, in other words, to be people at all" (Applebaum 2003, 335). The most extreme descriptions of hunger in Baltic women's deportation narratives make visible the awareness of the authors of the intentionality of the Soviet authorities in deliberately reducing people to such a state. Yet they also demonstrate the desperate efforts of the deportees to resist being dehumanized that are visible in attempts at keeping up at least some form of daily routine, in acts of caring, and, above all, in sustaining their moral strength that is instrumental in coping with even the most horrible suffering and loss caused by starvation.

Conclusion

Above all, the experience of deportation that constitutes an important aspect of the Baltic memorial culture that took shape during the period of regaining independence in the late 1980s is epitomized in terms of horrible suffering, martyrdom, bravery, and (heroic) survival on both individual and national scale (Hinrikus 2004, 63; Kirss 2004, 14). The memory of hunger, in particular witnessing family members and fellow deportees succumbing to it, emerges as one of the most traumatic aspects of the experience of deportation in many narratives. As I have argued in my previous research on the deportation and Gulag narratives of Baltic women, although most narratives contain ample evidence of potentially traumatic events and situations, testifying to the violence, injury, harm, and prolonged presence of fear and insecurity caused by the deportation (Kurvet-Käosaar 2012, 93), the narratives do not lend themselves easily to be read within the theoretical framework of trauma.

An analysis of the deportation narratives of Baltic women from the perspective of thematization of food makes visible a great number of different means and practices successfully employed to cope with food shortage and hunger and as such, constituting important practical as well of emotional strategies of survival. The analytical perspective taken highlights the key role of the continuities of everyday life that in turn shape both the representation of the gravity of suffering and the different means of coping. This approach also makes visible the significant role of creating a habitable day-to-day life, as well as developing and maintaining a communal attachment that facilitated survival in various small, yet indispensable ways. As the narratives demonstrate, despite the many grave deprivations of daily life, the deportees were able to build, and were not completely devoid of states of enjoyment, which in the narratives were frequently linked to food. There is also a close link between food and the development and upkeep of communal attachments, both imagined and real; this is an important source of moral strength and thus constitutes a vital condition for survival. These instances emerge as an important thematic line in the deportation narratives, and so emerging as the lighter, more hopeful side of the experience that needs to be weighed against and alongside the experience of traumatic loss and suffering in the consideration of the representation of the repressions of the Stalinist regime.

Disclosure Statement

No potential conflict of interest was reported by the author.

Funding

The article was written with the support of the Estonian Science Foundation grants [grant number ETF 9035], [grant number ETF 8875]; and Institutional Research Funding project Formal and Informal Networks of Literature, Based on Sources of Cultural History [grant number IUT22-2].

Notes

1. Issues relating to food can also be found in the narratives of men, although they do not form an equally strong thematic core in comparison with women's narratives. An important difference here to be taken into account is the fact that most narratives of men focus on the experience of forced-labor camp where the economies of food were different from those found in narratives of forced relocation. In the camps, meals (if this word would apply to the extremely scant food below any human standards) were provided by the *stolovaya*, or the dining hall (Applebaum 2003, 206). Compared with the deportees, camp inmates had fewer possibilities to improve their situation by the means of, for example, bartering their belongings, pinching, or using edible plants for cooking makeshift meals. Suffering from extended periods of hunger comes up in quite a few narratives, though more via factual mention than in a more elaborate thematic focus. Several narratives highlight the erosion of humanity as a result of dreadful living conditions, crushing workload and starvation. "Hunger dominated [our lives]" writes Igor Kullamaa, an inmate in a forced-labor camp near Magadan (Kullamaa 2000, 414). "In addition to hunger for food," he continues, "There was a different kind of hunger … a spiritual kind of hunger. Life that had as its leading morale 'You die today, I'll die tomorrow' excluded any humanness" (2000, 414). The narratives of men who were deported together with their mothers or other female narratives as children or adolescents form a separate category with respect to the thematization of food. Such narratives often discuss at length and with uttermost gratitude the efforts of their mothers to run the household and to feed their family well enough to ensure their survival. "Tribute must be paid to the women and mothers with children" writes Aleksandrs Birznieks, deported at the age of 21 and continues: "What the mothers did was nothing short of heroic, because many children did return home" (1999, 303).
2. Erna Nagel came from a wealthy rural background near Tartu, and according to her diary was at the time of her deportation primarily engaged in her studies at the University of Tartu along with an active social life; she was not involved in any household chores. Although the entries focusing on her life in Bondarka show that her mother did most of the cooking, the diary also shows that she had at least some cooking skills and could, for example, prepare a festive meal for her mother's birthday or for other holidays. Cooking skills of other women whose narratives have been the basis of the current study depend on their social background, but as importantly on their age and familial status: those women, who were married most probably were responsible for cooking back at home and had some culinary skills. Others who were children or adolescents, learned how to cook within the community. Most, though not all narratives, contain at least some information about the social background of the authors, though few if any contain details concerning the culinary traditions at home. Erna Nagel's diary, which offers a daily record of her life in the village of Bondarka in the

Siberian taiga, forms a notable exception, as procuring, preparing, and consumption of food constitute the main thematic focus of her diary.

References

Anušauskas, A. 2013. Deportations of 14–18 June 1941. Accessed August 10. http://www.komisija.lt/Files/www.komisija.lt/File/Tyrimu_baze/I%20Soviet%20okupac%20Nusikalt%20aneksav/Tremimai/ENG/1941_BIRZELIO_TREMIMAI-EN.pdf.

Applebaum, A. 2003. *Gulag: A History*. New York: Doubleday.

Birznieks, A. 1999. "I Was Twenty-One." In *We Sang Through Tears: Stories of Survival in Siberia*, edited by A. Sics, 285–308. Riga: Jānis Rose Publishers.

Bīviņa, L. 1999. "Winter by the White Sea." In *We Sang Through Tears: Stories of Survival in Siberia*, edited by A. Sics, 109–128. Riga: Jānis Rose Publishers.

Budrytė, D. 2010. "Experiences of Collective Trauma and Political Activism: A Study of Women 'Agents of Memory' in Post-Soviet Lithuania." *Journal of Baltic Studies* 41 (3): 331–350. doi:10.1080/01629778.2010.498191.

Caruth, C. 1996. *Unclaimed Experience: Trauma, Narrative, and History*. Baltimore: Johns Hopkins University Press.

de Silva, C. 1996. "Introduction." In *In Memory's Kitchen. A Legacy from the Women of Terezín*, edited by C. de Silva, xxv–xviii. Lanham: Rowan and Littlefield.

Geka, D. 2012. "Introduction." In *The Children of Siberia*, edited by D. Geka. Vol. 7. Riga: Fonds Sibīrijas Bērni.

Grinkevičiūtė, D. 2002. *A Stolen Youth, A Stolen Homeland. Memoirs*. Vilnius: Lithuanian Writers' Union Publishers.

Hinrikus, R. 2004. "Deportation, Siberia, Suffering, Love. The Story of Heli." In *She Who Remembers Survives: Interpreting Estonian Women's Post-Soviet Life Stories*, edited by T. Kirss, E. Kõresaar, and L. Marju, 62–77. Tartu: Tartu University Press.

Hirsch, M., and L. Spitzer. 2006. "Testimonial Objects: Memory, Gender, and Transmission." *Poetics Today* 27 (2): 353–383. doi:10.1215/03335372-2005-008.

Jolluck, K. R. 2002. *Exile and Identity. Polish Women in the Soviet Union during World War II*. Pittsburgh: University of Pittsburgh Press.

Kaber, M. 2000. "Elulugu." In *Eesti Rahva Elulood II*, edited by R. Hinrikus, 146–157. Tallinn: Tänapäev.

Kalēja, V. 1999. "Memories." In *We Sang Through Tears: Stories of Survival in Siberia*, edited by A. Sics, 19–58. Riga: Jānis Rose Publishers.

Kaļiņina, H. 1999. "We Sang Through Tears." In *We Sang Through Tears: Stories of Survival in Siberia*, edited by A. Sics, 71–88. Riga: Jānis Rose Publishers.

Kirss, T. 1999. "Ariadne Lõngakera: Eesti Naiste Siberilood Nende Elulugudes." In *Paar Sammukest 16: Eesti Kirjandusmuuseumi Aastaraamat*, edited by S. Olesk, 23–31. Tartu: Eesti Kirjandusmuuseum.

Kirss, T. 2004. "Introduction." In *She Who Remembers Survives*, edited by T. Kirss, E. Kõresaar, and M. Lauristin, 13–18. Tartu: Tartu University Press.

Kirss, T. 2005. "Survivorship and the Eastern Exile: Estonian Women's Life Narratives of the 1941 and 1949 Siberian Deportations." *Journal of Baltic Studies* 36 (1): 13–38. doi:10.1080/01629770400000221.

Kullamaa, I. 2000. "Elulugu." In *Eesti Rahva Elulood I*, edited by R. Hinrikus, 410–419. Tallinn: Tänapäev.

Kurvet-Käosaar, L. 2005. "Imagining a Hospitable Community in the Deportation and Emigration Narratives of Baltic Women." In *Women's Life-Writing and Imagined Communities*, edited by C. Huff, 59–78. London: Routledge.

Kurvet-Käosaar, L. 2012. "Vulnerable Scriptings: Approaching the Hurtfulness of the Repressions of Stalinist Regime in the Life-writings of Baltic Women." In *Gender and Trauma. Interdisciplinary Dialogues*, edited by F. Festic, 89–111. Newcastle upon Tyne: Cambridge Scholars Publishing.

Kurvet-Käosaar, L. 2013. "Voicing Trauma in the Deportation Narratives of Baltic Women." In *Haunted Narratives: Life Writing in an Age of Trauma*, edited by G. Rippl, P. Schweighauser, T. Kirss, M. Sutrop, and T. Steffen, 129–51. Toronto: Toronto University Press.

Laar, M. 1990. *14. Juuni 1941. Mälestusi ja Dokumente*. Tallinn: Valgus.

"Lapsena Külmal Maal." 1990. In *14. Juuni 1941. Mälestusi Ja Dokumente*, edited by M. Laar, 25–29. Tallinn: Valgus.

Lazda, M. 2005. "Women, Nation, and Survival: Latvian Women in Siberia 1941–1957." *Journal of Baltic Studies* 36 (1): 1–12. doi:10.1080/01629770400000211.

Long, J. 1999. *Telling Women's Lives. Subject/Narrator/Reader/Text*. New York: New York University Press.

Mertelsman, O., and A. Rahi-Tamm. 2009. "Soviet Mass Violence in Estonia Revisited." *Journal of Genocide Research* 11 (2–3): 307–22. doi:10.1080/14623520903119001.

Montanari, M. 2006. *Food Is Culture*. New York: Columbia University Press.

Nagel, E. 2007. *Erna Nageli Päevik. Olen Kui Päike Ja Tuul*. Tartu: Hotpress.

Rahi-Tamm, A. 2007. "Deportations in Estonia, 1941–1951." In *Soviet Deportations in Estonia: Impact and Legacy*, edited by K. Kukk, and T. Raun, 9–54. Tartu: Filiae Patriae Society.

Sink, P. 1999. "Kannatuste Rada." In *Me Tulime Tagasi*, edited by R. Hinrikus, 32–71. Tartu: Eesti Kirjandusmuuseum.

Skultans, V. 1997. *The Testimony of Lives: Narrative and Memory in Post-Soviet Latvia*. London: Routledge.

Tallo, E. 1999. "Minu Mälestused." In *Me Tulime Tagasi*, edited by R. Hinrikus, 140–76. Tartu: Eesti Kirjandusmuuseum.

Todorov, T. 1996. *Facing the Extreme: Moral Life in the Concentration Camps*. New York: Henry Holt and Company.

Vilnis, L. 1999. "The Dark Pages of My Life." In *We Sang Through Tears: Stories of Survival in Siberia*, edited by A. Sics, 89–108. Riga: Jānis Rose Publishers.

THE EVOLUTION OF HOUSEHOLD FOODSCAPES OVER TWO DECADES OF TRANSITION IN LATVIA

Lani Trenouth and Talis Tisenkopfs

This article traces changes in household food consumption patterns comparing the late Soviet period and the present day based on household interviews and interpretive analysis. We reconstruct and visualize four foodscapes from households of varying socioeconomic contexts, based on their memories of food consumption prior to the breakup of the Soviet Union and their lived experiences of food consumption today. These vignettes are a basis to discuss changes in food consumption patterns in the broader context of transition. This study aims to contribute to a greater understanding of the evolution of food consumption from the perspective of the everyday consumer.

Human food consumption practices are highly contextual and continuously changing. They are shaped by macro level political, economic, and cultural structures, transformed by national and transnational level policy decisions, technological developments, and market trends, and ultimately articulated through personal values, knowledge, and access to various forms of social and economic capital. The major transformations seen in some ex-Soviet countries from 1991 onwards offer fertile ground for exploring changes in everyday food consumption practices of "regular folk." In this article, we take the case of Latvia as an example of a post-socialist[1] country to explore and describe the evolution of household foodscapes from the perspective of consumers over the last two decades of transition. We apply the foodscape concept to exploration of food habits, in particular the long-term changes in household food consumption influenced by the political change and market transition. Our main research question addresses the processes, factors, and agents of reconfiguration of household foodscapes as an effect of wider food system transformation. We address this question by asking what practices have been added or removed from the repertoire of everyday consumers and concomitantly in what way have other practices evolved in their significance. We are interested not only in new household foodscape agents and relationships emerging from the mainstream industrialization and concentration tendencies in agri-food chains, but also in longevity and reshaping of traditional food consumption practices and emergence of new "alternative" food relationships as a part of dynamic foodscapes.

The notion of foodscapes is increasingly used within food studies, public health, and nutrition science as a tool to describe how food, places and people are interconnected and how they interact (Mikkelsen 2011). Mikkelsen distinguished three levels of foodscapes: macro institutional, mezzo or community level and micro level foodscapes of household food habits. Brembeck and Johansson (2010) defined foodscapes as signifying all the places where one comes into contact with food and eating. The term foodscape has been used also to distinguish the dispersed and fluid aspects of food distribution and consumption, for example ethical dimension of organic, local, and fair-trade supply chains (Morgan 2010). Ethnologists have used the term "mental foodscape" to discuss the places of eating out (Bildtgård 2009). In tourism studies, the notion of foodscapes has been used to analyze imaginary geographies of food related travel destinations. Jakob Wenzer (2010) has applied foodscapes concept to the analysis of new food consumption sites in cities. Sociologists have used the foodscape lens from a perspective informed by political science to highlight the occupying of retail space by large companies focusing on the availability of food stores, and on disparities in food access along the lines of neighborhood, race, and income (Filomena, Scanlin, and Morland 2013). Focusing on foodscapes emphasizes the social, relational, and political construction of food thus highlighting not simply food provision but also questioning the existing power structures causing inequalities in access to food. The notion has been used in critical geographical research into urban food poverty (Miewald and McCann 2014). Thus, the foodscape concept has proved to be a useful analytical tool and also a vehicle to build strategies and actions for more sustainable food provision. Hansen and Kristensen (2013) have demonstrated how creation of institutional foodscapes around school food programs has positively influenced consumption habits and nutrition of children.

Our research focuses on micro-level household foodscapes that include also elements of wider economic, geographical, institutional, and cultural characteristics of food system. For the purposes of this article, we understand the foodscape to be the contextual milieu in which food consumption takes place, including social, cultural, economic, political, and geographic spaces. At the household level, foodscapes are constructed in the articulation of meaningful, culturally, and socially determined activities of food consumption that take place in a wider economic and political context.

Food consumption as a subject of academic interest in post-socialist Europe has not received the same degree of sustained attention as seen in some western European countries. This regional disparity appears to be changing over the last decade, with a number of ethnographic and a handful of quantitative studies being published in journals and edited volumes. Much of the new scholarship, which is focused on food consumption in post-socialist countries, uses food as a lens to view other topics of societal and academic interest, uses ethnographic methods, and has been carried out in Russia (cf. Patico 2002; Shevchenko 2002; Caldwell 2002, 2004; Ries 2009; Yamin-Pasternak 2008). Potatoes, chocolate and cognac, and "Soviet" sausages are used as entry points to explore contradictory Russian cultural icons (Ries 2009); symbolism in gifts within social networks in Russia (Patico 2002); gender and food roles in Hungary (Fischer 2010); and nostalgia and a post-Soviet hybrid modernity in Lithuania (Klumbytė 2010).

A number of scholars have taken the phenomena of the *dacha* (household garden) as their focus of interest, exploring the meaning of the practice through a variety of approaches (cf. Rose and Tikhomirov 1993; Seeth et al. 1998; Hervouet 2003; Zavisca 2003; Alber and Kohler 2008; Round, Williams, and Rodgers 2010; Caldwell 2011; Jehlička and Smith 2011; Jehlička, Kostelecký, and Smith 2013). These studies did not directly address the topic of food consumption, although some conclusions on consumption could be inferred from their work. Some of the attitudes and beliefs toward household food production as described in some of the more ethnographic works spillover into the realm of consumption. For instance, a motivation to grow one's own food based on the desire for organically produced, chemical free foods reasonably indicates a concomitant desire to *consume* such foods.

A further and smaller body of work on consumer behavior and marketing has also been carried out (cf. Banyte, Brazioniene, and Gadeikiene 2010; Horská, Ürgeová, and Renata 2011; Zagata 2012). These works have contributed rich details and insights, however the field of food studies in general, and food consumption in particular, in the post-socialist context, remain rather underexplored, theoretically and empirically.

In this study, we have chosen to use a life story approach to qualitatively explore the cultural and socioeconomic phenomena surrounding food consumption practices in Latvia today and 20 years ago, where respondents narrate their experiences and memories of the everyday practice of food consumption. From these narrations, we ask: in what ways has their foodscape changed over the last 20 years? What practices remain, what practices have been lost, what new practices have emerged? Of those practices that have remained, how have they changed in their importance and relevance over the last 20 years? Have some practices remained but their meanings and motivations evolved over the years?

Our method is reconstructive analysis, comparison, and visualization of foodscapes derived from life story interviews with a number of socially and economically typical households. Taking these interpretive vignettes, along with additional information from other interviews, we draw upon parallels and tensions between Latvia and other post-socialist contexts to explore what these changes in foodscapes mean and broadly comment on changes in patterns, motivations, and meanings of food consumption in Latvia. Our narrative and visual mapping of these household foodscapes in transition illuminates some striking shifts in food habits over the last two decades, notably the disappearance of certain food practices, the continuation of others, and the emergence of some entirely new habits and relationships. Our research supports the idea that current food practices are only partially mediated by the socialist experience and partially by new realities. There are some current food habits that are embedded in older traditions of the pre-socialist period, as well as new behaviors and routines that have emerged completely, and still others that echo the practices established in recent history.

Approach and Methodology

Taking an actor-centered constructivist approach, we (re)construct narratives of the foodscapes of the late Soviet period and of today. A total of 17 persons were interviewed, either individually or in groups, over a period of two months in early 2012. Interviews ranged from one hour to two and a half hours each. Respondents were between 23 and 74 years of age, and they were from varying educational and occupational backgrounds and income brackets, with a slight majority being women. Consumer households were recruited through a snowball technique using multiple entry points. We were seeking diversity in social *habitus* (Bourdieu 1984) rather than representativeness in our selection of households in order to explore food consumption experiences across diverse social groups. The households were selected based on socioeconomic divisions and processes in Latvian society: an urban professional household (the rising middle class), a rural household (rural restructuring), a working class household (effects of recent economic crisis), and a pensioner household (aging of the population). Most interviews were carried out in a small town in central Latvia, with some interviews conducted in the capital.

Using an inductive grounded theory approach (Glaser and Strauss 1967; Goulding 2002), we iteratively identified, coded, and categorized emergent themes from the interviews. We adopted the constructivist, agent-based approach of actor-network theory (ANT) (Latour 2005; Law and Hassard 1999) in our depiction of household foodscapes. ANT reflects material-semiotic reasoning whereby nonhuman objects and concepts also possess agency within social networks and their relationships can be mapped. Such network "actants" discussed by our respondents included objects such as refrigerator or foreign foods, and concepts such as organic or convenience. The foodscapes were visualized using the tool Smart Network Analyzer (Krebs 2000) whereby the various actants within the respondents' foodscapes became the nodes in the foodscape map.

The development of the foodscape maps posed some challenges in defining exclusive thematic clusters using the interview data. This was because some foodscape actants simultaneously hold multiple meanings for the same respondent and among respondents, and because some thematic categories cannot be clearly disentangled from one another. This article presents the results of only a few household interviews, customary with interpretive work, we therefore do not claim representativeness in our analysis, rather our work provides illustrative details as a basis for further research.

Before turning toward the foodscape vignettes, a methodological caveat bears discussion here, as a large part of this study rests on our respondents' recollections. Memory is rather selective and pliable in nature and can be swayed by broader collective social narratives of the past as well as how the remembered past is positioned in contrast with the experienced present. While it is not our intent to question the fidelity of the image of the past as remembered and described by our respondents, it would be prudent to be aware of these issues concerning memory.[2]

Foodscapes of "Then" and of "Now"

The following four household vignettes illustrate the experience of foodscape changes over the last 20 years, exploring what was "then" and what is "now." As a shorthand, we refer to these households as *urban professional, rural farming, urban retired*, and *urban working*, with each household embodying variable socio-economic characteristics. Respondents framed their narratives of food consumption according to the specifics of their everyday life and personal memories of transition. Themes of knowledge, choice, modes of acquisition (buying, growing, gleaning), food preparation, and eating emerged naturally from the life stories given by the interviewees and are reflected in the accompanying figures for each household. What follows is an illustrative interpretation in the form of text and visual foodscape maps generated out of the narratives told by each household comparing past and present concerning these themes.

Urban Professional

This household consists of a married couple in their 40s living in the capital city, Riga. The husband works as a professional artist and the wife is a manager in the public sector. The couple's lifestyle is typical of the new urban middle class and is characterized in large part by recent upward mobility. Leisure activities figure prominently and travel abroad has become relatively commonplace. Career demands are high and the couple increasingly faces time constraints in their day-to-day lives. They have readily embraced the wider diversity of consumer choices offered by the new market economy.

The narrative of this household clearly highlighted the significant increase in the range of elements considered when making food choices, whereas in the past "*we bought whatever was available.*" Among the most relevant elements are taste, fashion, local foods, organic, and travel. None of the factors described by the household as playing a role in their decision making today were also described as being important in the past (Figure 1).

It is significant that for this household neither growing food nor social exchange networks are particularly relevant practices either in the past or the present. The range of sources for purchasing food has increased, with supermarkets and the night market appearing as new providers. For this household, affordability, availability, and diversity of food have all increased. The appearance of modern kitchens and kitchen appliances, including a freezer, has influenced the purchasing pattern as well as preparation practices of the household. Time constraints have dramatically increased

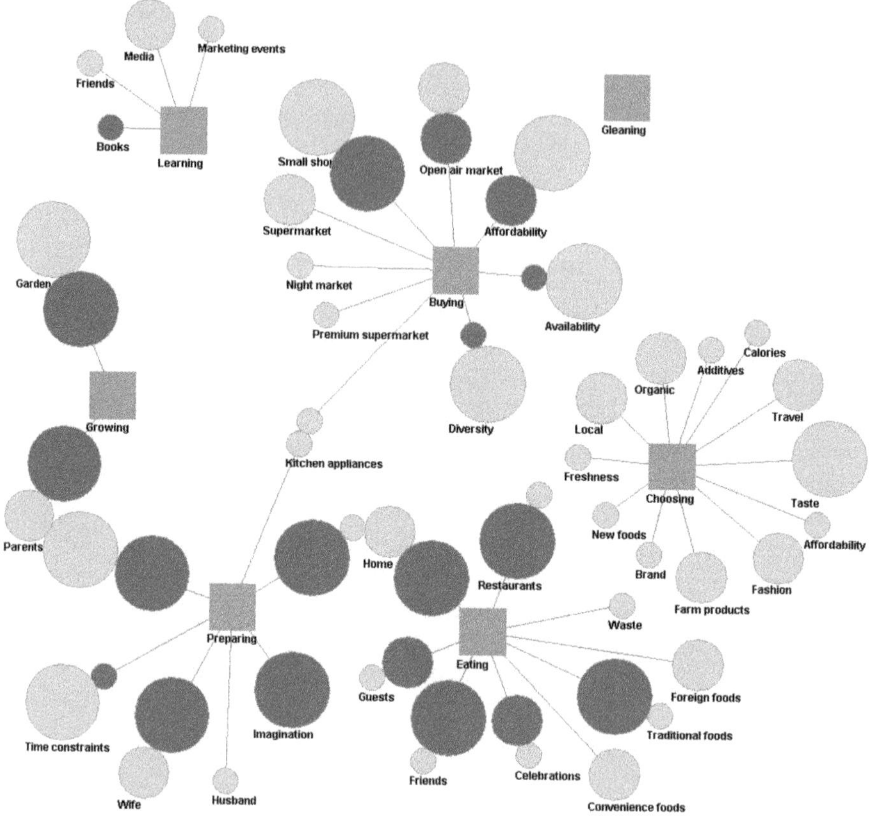

FIGURE 1 Foodscape network map, past and present: Urban Professional.[3]

and home cooking has decreased. Food is increasingly eaten outside the home with convenience foods, foreign menus, and exotic foods playing a role today, which was absent in the past.

The household described the importance of creativity in food preparation in the past that was a direct result of the limited ingredient options available, a situation that is no longer relevant today. Books, friends, media, marketing campaigns, and travel now play a role in shaping food preference, and sophistication is held in higher esteem than tradition. The social function of eating for this household has decreased drastically likely as a function of general individualization of the society and changing norms of socialization. Where formerly it was commonplace to receive friends for dinner, today this practice is less common, and requires a great deal more planning and preparation. Today, this household is actively considering taste preferences, value orientations, purchasing power, food knowledge, and representation of social status at occasional celebrations and meals with friends.

This household's foodscape today can be characterized by an increase in food diversity, rise in importance of convenience, and greater functional performance of food in social communication and status recognition.

Household Vignette: Rural Farming

This rural farming household consists of two parents in their 40s and three children aged 13–16. Since the late 1980s this household has lived on a 20 hectare farm practicing small-scale agriculture. In addition to farming, the husband works in a forestry company in a nearby village and the wife has no paid employment. Unable to meet modernization requirements in order to integrate into EU industrial agriculture systems, the couple decided to move away from an "agri-culture" mode toward an "eco-culture" mode of production. The farm is not certified organic but most production practices are characterized by the owners as "deliberately environmental." The couple is seeking out new innovative possibilities such as tapping into markets for niche products such as organic, free-range eggs (Figure 2).

Clearly evident in the narrative of this rural farming household is the importance of both self-provisioning and of ethical production practices. This household is organized around the philosophy and practice of sustainable self-provisioning. Choices are determined by values of health, product safety, animal wellbeing, and care for nature. Beyond typical fresh garden products, this household provides itself with a wide diversity of traditional foods including forest fruits, juices, pickled/ preserved foods, meat, and sausages. On average, the family produces about half of

FIGURE 2 Foodscape network map, past and present: Rural Farming.

the food they consume. At the same time, growing food is increasingly costly and time constraints have become more of an issue. Sporadic purchases connect this household foodscape to diverse retail structures. The opening of supermarkets in the area has meant that the household visits small local shops with decreasing regularity in favor of supermarkets.

The significance of social networks of relatives, neighbors, and acquaintances is greater now than in the past, and surpluses from farm production are sold to relatives and friends. Gathering wild food has been described as occurring equally in the past and present, while hunting is a new practice, although neither was described as highly significant. Home food preparation has not noticeably changed over the last 20 years. Eating out today is related to travel or occasional visits to restaurants, whereas in the Soviet period it was limited to workplace canteens.

Factors influencing food choice for this household have increased significantly over the last two decades. Concerns about food additives, animal welfare, chemical residues on produce such as pesticides or fertilizers, food miles, and health have all emerged recently. Considerations related to organic and traditional foods, while present in the past, figure more prominently today. Avoidance of highly processed, convenience, and industrial foods is based on convictions of their low nutritional quality and negative health impacts.

This household's foodscape today can be characterized by a concern about holistic health (human, animal, and environment), an attempt to increase self-reliance in food provisioning, and an expanding social network for exchanging foods.

Household Vignette: Urban Retired

This household consists of an elderly couple living in a compact two-room apartment in a building on the edge of a small town. In a common story of outmigration from Latvia, four out of their five children from former marriages have moved to Ireland. Previously both were engaged in the workforce and today the couple lives on modest pensions, which barely allow them to make ends meet. Despite these challenges, the household does not consider themselves to be "poor," indicating that they are not at risk for eviction due to unpaid debts and they are able to provide themselves with three meals a day (Figure 3).

The most striking element of this households' foodscape is their high dependence on purchasing low cost, typically industrial, foods. Price and available discounts on food items figure prominently in their food choices. Concerns about additives, chemical residue, and trust were also elements mentioned as being relevant today, but their lack of purchasing power limits their ability to access foods they consider to be healthier.

In the past, each of the household members acquired much of their food through collective purchase of farm products organized by their employee trade unions, and other items bought at small shops. Today, purchasing is done predominantly at supermarkets and occasionally supplemented through informal purchase from Roma who offer wild foods such as berries and mushrooms. When describing their buying habits, the process of careful price comparison across supermarkets was described in

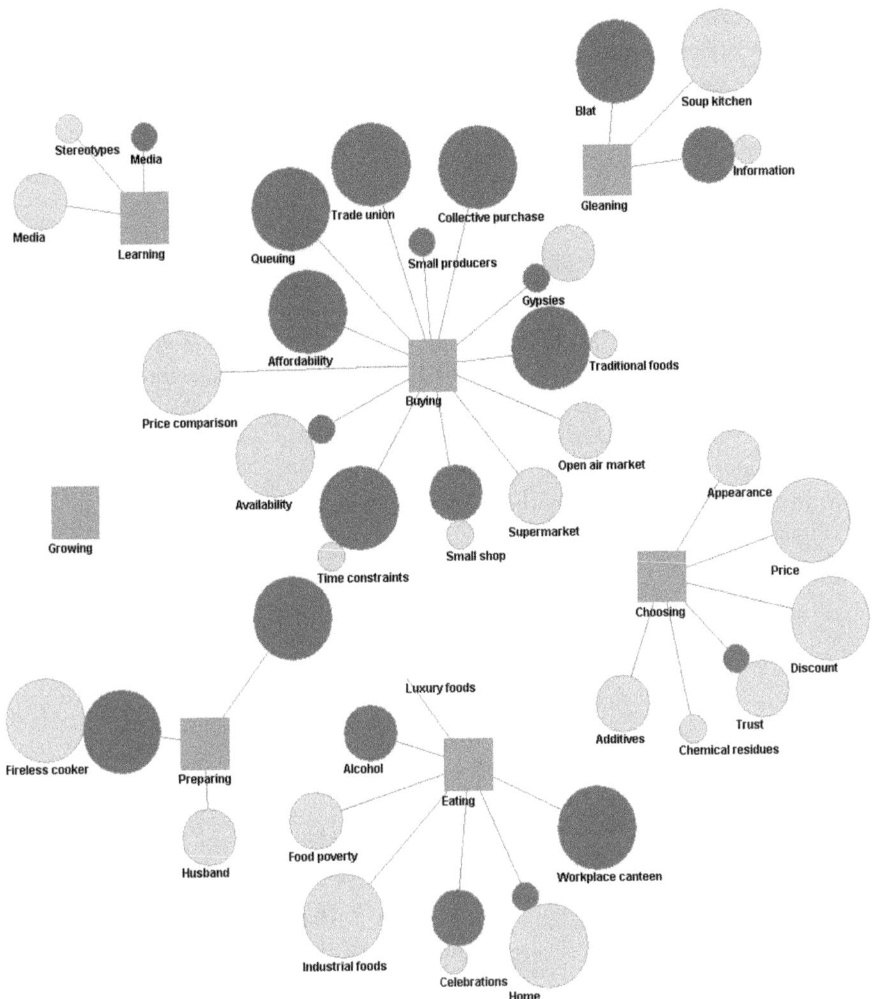

FIGURE 3 Foodscape network map, past and present: Urban Retired.

detail, and food from the open air market was deemed unaffordable. The household revealed that affordability and queuing are no longer present in their foodscape and that time constraints and traditional foods figure less prominently as well.

This household does not grow any food themselves nor do they engage in any food exchange through social networks, although the husband makes regular trips to the town soup kitchen, where a liter of soup is given to take home. Formerly, the workplace canteen played a significant role in their eating practices, but they now eat virtually all of their meals at home. Cheap, industrial foods figure more prominently today and luxury foods have never existed in their foodscape, either in the past or the present. While the household is able to secure three meals a day, there is an element of food poverty, especially in the nutritional quality of the foods they eat.

This household's foodscape today can be characterized by the sense of unaffordable food today, nostalgia for the Soviet period, and the almost total absence of a social exchange network.

Household Vignette: Urban Working

This family, a couple in their 50s, lives in the center of a small town. Both of their two children are married, and one has left for England. In a familiar story of the post-socialist transition period, the husband lost his job when the construction company where he worked closed business. Since then, he has worked as a freelance construction worker in private housing projects. Income was consistent and reasonable during the economic boom of 2004–2007 after Latvia joined the European Union, but dropped considerably during the recent economic crisis. The husband is the only breadwinner in the household as the wife finds it difficult to secure regular employment (Figure 4).

FIGURE 4 Foodscape network map, past and present: Urban Working.

Many elements, which figure prominently in food selection today, such as freshness, naturalness, organic, smell, and additives, were also concerns for this household in the past. Health and dietary concerns, while also relevant in the past are even more prominent today. New factors in food choice include solidarity with farmers and the provenance of the food, with local being preferred. Frozen food has never factored into this household's diet.

The wife of the household has a history of actively learning about food, and books, and learning by doing continue to be important ways of learning for her. More recently the Internet and travel have increased their food knowledge in new ways. The household highlighted the importance of transmitting what they learned about food to their children and grandchildren.

The diversity of sources from which food is acquired has increased, with supermarkets and specialty shops among the new sources. The importance of open air markets and buying from farmers in these markets has increased for this household, as has the purchase of exotic and foreign foods from supermarkets. Traditional foods continue to play an important role while convenience foods, on the other hand, have not made entry into the foodscape of this household. Queuing, while remembered as significant in the past, is no longer a daily experience.

This household's reliance on growing their own food is relatively unchanged from the past to the present, but the recent economic downturn has led to an increased motivation to garden owing to the higher price of food. Food exchange through social networks does not appear strongly, although relatives do share land for gardening. Wild foods are still present in their foodscape but remain as a minor addition. Use of *blat* (an "economy of favors," see Ledeneva 2009) as a means to acquire food has disappeared.

Home cooked foods are prepared daily by the wife of the household, much as it was in the past. Eating is more often done in the home setting and the family continues to play an important role. However, since the children have left home and her husband travels for work from time to time, the wife eats alone more often than previously. The diversity of foods eaten has increased and more meals now occur in restaurants.

This household's foodscape today can be characterized by complexity and reflexivity in food decisions, overall balance, and relative continuity from the past to the present and on into the future.

Foodscapes in Transition: Changing Patterns, Motivations, and Meanings of Food Consumption

We now turn to some of the broader patterns, motivations and meanings behind the foodscape practices described above in order to enlarge our understanding of the evolution of food consumption practices from the perspective of the everyday consumer through the memories of their changing behaviors. We focus on the elements important in choosing, buying, growing, and exchanging food, as these were most prominent in the stories they shared. Again, we include a visual interpretive representation of the magnitude of change of each actant when comparing past and present

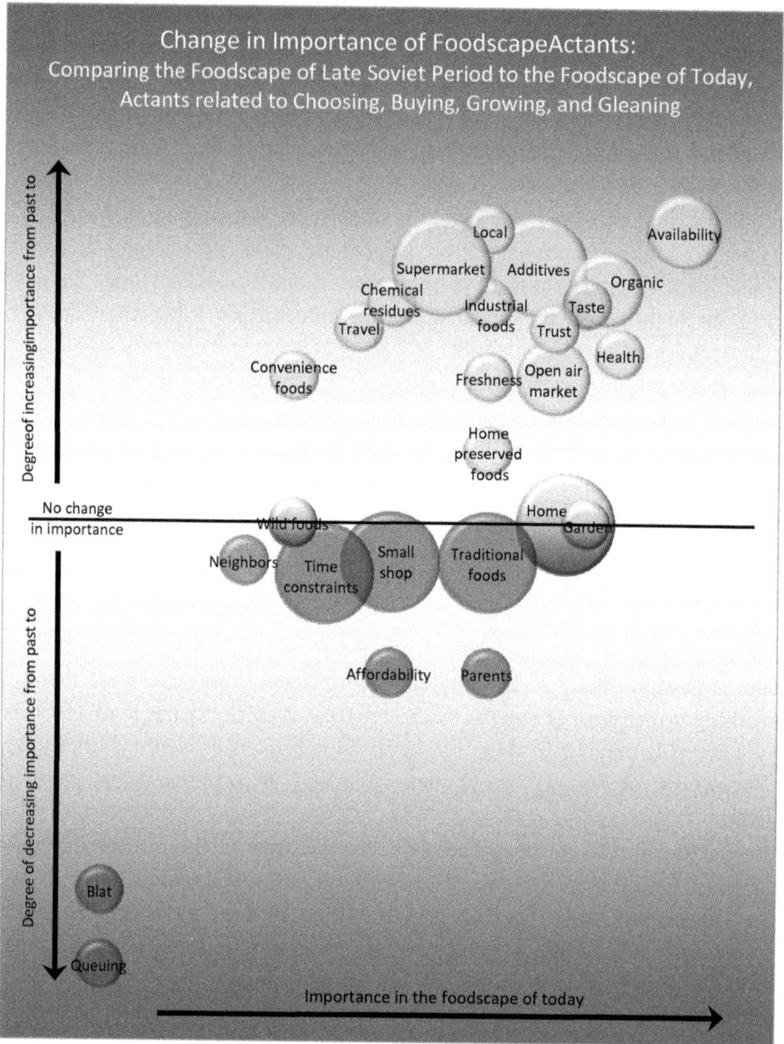

FIGURE 5 Change of importance of foodscape actants.[4]

foodscapes, and the importance of the actant on the foodscape of today. This graphic represents all four households profiled above as a composite and gives a broad sense of what elements are important today, the relative degree of their importance, and how this has changed from the past (Figure 5).

Choosing

> We went through a period of eager experimentation with the new, foreign foods but quickly turned back to the foods we know ... they just taste better. (Anonymous, March 2011)

When the borders opened up to new foreign products in the early 1990s, there was a sense of curiosity and excitement surrounding the new foods suddenly available. After a short period of experimentation, however, many described returning to the foods they were familiar with, citing their disappointment in the artificiality and lack of taste of many of the new foods. Among many consumers, there is a common discontent with what Buchler, Smith, and Lawrence (2010) term "modern risks" in the form of increased adulteration of foods, presence of chemical residues from pesticides and fertilizers, and the addition of artificial additives. In contrast, food produced in the Soviet period is remembered as less processed, closer to its natural form, and grown with less application of synthetic chemicals. Many Latvians from all socioeconomic groups continue to value these characteristics and now find themselves actively seeking out such foods amongst the flood of highly industrialized products now in the marketplace.

Taste was often cited as a key consideration in food choices, especially amongst men. In some cases childhood memories of taste figure prominently, and the desire to find the same taste in today's foods was mentioned. The degree and quality of taste appear to be strongly correlated with other characteristics such as food considered to be organic, natural, fresh, and traditional. Informants in our study often suggested that "Latvian foods" (*Latvijas produkti*) and "local food" (*vietējie produkti*), terms which were often used interchangeably, were more likely to embody these characteristics and for that reason they were preferred. Simply put, informants felt that organic, clean, natural, Latvian products, reminiscent of what they ate as children, simply tasted better than their more adulterated, new, and foreign counterparts. These sentiments reinforce Schwartz's (2007) suggestion of a relationship between nature and national identity in Latvia, and parallels findings of Caldwell (2011) in her studies of Russian *dacha* culture.

Health concerns have become more prominent in today's decision-making process, especially amongst women, where "healthy food" is often associated with organic, unadulterated products, relatively free of additives and chemical residue from production. Šūmane (2010) noted that in the initial phase of organic agriculture in Latvia in the 1990s, health concerns were among the driving factors for farmers in their transition to organic agriculture.

Offsetting health concerns are worries over the cost of food that loom larger today than in the past as a result of the greater reliance on a cash-based commodity market where food access is largely dependent on an adequate income. Informants in this study described the ways in which they try to balance cost with other considerations and preferences. For instance, some grow their own organic produce and others carefully monitor supermarket flyers for special offers. The emergence of new charitable food programs and the reported rise in demand for such services speaks to the difficulty faced by some households in meeting their minimum food needs – a situation which our informants describe as being a new phenomenon.

Buying

> Before, you had money but there was nothing to buy, now there are things to buy but there is no money. (Anonymous, March 2011)

While the food regime of Latvia is increasingly dominated by industry and retail, it should be recalled that monetized modes of food acquisition have long played a significant role, regardless of the prevailing type of economic system. This is especially so for urban residents lacking adequate means to engage in self-provisioning in any substantial way. In the past, options were often limited to a relatively small range of items on offer in the formal economy, what could be self-produced, or what could be acquired through social connections. Smith (2003, 182) characterized shopping in Hungary under the Soviet economic system as a passive activity whereby the shopper could sense where goods would be on offer and queue for hours looking forward to whatever would be available and he notes that "the citizen asks not 'What shall I buy today?' but 'What are they giving us today?' ". This form of food acquisition is no longer seen in Latvia, and while there is generally an appreciation of the wider variety of items available today, consumers are now faced with the task of making decisions in a far more complex choice environment.

Respondents cite a much stronger dependence on formalized monetary market systems and less on home production practices and social networks. The role and importance of monetary exchange has increased significantly as a result of this change in the economic system, and Smith (2003, 187) suggested that money has acquired "a new fetishized power." Liberalization of markets has brought with it the challenge of acquiring inexpensive yet quality foods (Patico and Caldwell 2002). Many of our informants confirmed this by explaining that, unlike in the past, in today's society almost anything can be acquired – so long as you have enough money.

We are reminded that socioeconomic class determines what resources people have access to which enables them to develop certain preferences and to enact them, and where "positions in the socioeconomic hierarchy are made apparent in their relationship to the means of consumption" (Eglitis 2011, 426). For some consumers, new buying and consumption actants such as supermarkets, premium supermarkets, specialty shops, restaurants, takeout facilities, exotic foods, and others have emerged for consumers to exercise their choices and perform food lifestyles along the dimensions of income, social status, values, and lifestyle expressions.

Growing

> Previously, if you didn't have money, at least you had your garden so it wasn't so bad. Now it's different, because for some people if you lose your job you really don't have anything. (Anonymous, March 2011)

> No matter what, you can rely on your garden to get by. (Anonymous, March 2011)

Paralleling other post-socialist contexts (Caldwell 2011; Smith 2002; Smith and Jehlička 2007), the prevalence of household food production in Latvia was, and continues to be, due to both economic and cultural/historical forces and not merely a coping strategy when faced with austerity. Household food production must be seen within a broader context where long-standing agrarian practices constitute the historical and cultural backdrop. As Alanen (2004) described, in the Soviet period there was a strong tradition of household farming in Baltic countries that existed alongside official collectivized agriculture and provided rural and peri-urban households with an important source of income as well as food.

By the same token, the gathering of wild foods is a widespread practice with strong pre-Soviet origins, which remains enormously important as a historical and cultural practice.

For many, gardening is further instilled with a sense of autonomy rather than economic necessity, as noted in Zavisca's comprehensive study in Russia (2003). By maintaining autonomy over some aspects of food acquisition, people are able to secure a measure of greater agency in the availability, variety, and quality of the foods available to them, and can offset any hardships related to loss of income. However, while still widely practiced, especially in comparison with western European countries, fewer households maintain a garden today and those who do are tending smaller plots of land and/or growing a much narrower range of produce. Household food production of vegetables, fruit, dairy, and eggs was an extremely common practice in the Soviet period, providing a stable and significant portion of the household consumption of these items. But, today the relative importance of gardening for subsistence or as an economic practice is in decline and is increasingly a matter of culture. Underscoring this point, Caldwell (2011, 82) observed sentiments of intimacy or even spirituality with nature and "nostalgia for foods that come from rural spaces." These sentiments were echoed by a number of our respondents.

Market failures of the socialist economic system led some people to seek a degree of self-sufficiency in food provision outside the state managed economic system. An economy of shortage no longer prevails, but a new form of market failure has emerged whereby people are unable to easily access acceptable food – for example, food that is natural, organic, and/or free of additives. For some of our informants gardening was a direct response to the inability to buy organic produce at an affordable price. Gardening may then represent, among other consumption activities, a new form of resistance directed at the increasing dominance of supermarket logic (Smith 2002).

Exchanging

> In Soviet times there was a greater need for more social networks, and now it is more impersonal, both with food and other necessary items. (Anonymous, March 2011)

Reciprocal exchange and gifting of food and other items or favors has often been cited as playing an important role in building and maintaining social capital across support networks in the socialist period (cf. Caldwell 2004; Ledeneva 2009). The need to develop and maintain good social connections in the Soviet period was a way to mediate economic risks (Mincytė 2014), and *blat*, an "economy of favors" (Ledeneva 2009), was called upon to gain access to items which were particularly difficult to come by through the use of special social network connections. In the past social capital was a critical resource for acquiring the items needed for day-to-day living, but in today's more capitalistic markets in Latvia the process of commodity exchange has become much more impersonal. While reciprocal exchange networks do remain relevant today, particularly for households in rural settings, its importance is in dramatic decline. Overall, there is a reduction in noncommoditized exchange through social networks and less buying through trusted personal connections. Younger people may be easily embracing this new form of market interaction, but for the older generation it can be bewildering and stressful (Fischer 2010).

Social networks today are smaller and more atomized both because of the prevailing social process of individualization and de-collectivization (Sīmane 2003) as well as a result of the significant outmigration of Latvians in their 20s and 30s in recent years. Despite the apparent process of dissolution of social networks, trust-based social relationships still play a role in household foodscapes. We found similar reasons to Caldwell (2002) for the preference of some to purchase food through social networks, including personal guarantees on the quality of the product, a means to complain if the product did not meet expectations, the ability to ask questions about the characteristics of the product and the expectation of receiving honest answers, all of which generated a greater sense of faith and trust in the product. Preserved foods, gifts from the garden, eggs, and even meat are still exchanged within families and among neighbors. Some households engage in the purchase of raw milk through networks of trust and many consumers have a strong preference for making their purchases in the market from farmers and vendors they know.

Trust and social capital were in decline during the early transition period, but appear to regain footing in some narrow networks. Sometimes inspired by foreign alternative consumption models, new relationships within and beyond economic transactions are opening up. Still, these collective and civic food networks in the context of post-socialism are fragile as an effect of weak institutional and interpersonal trust within the society. Consumer organizations in the food sector are virtually missing, and cooperative practices are weak both in the production and consumption realms of food (Tisenkopfs et al. 2011).

Conclusion

The household foodscapes explored in this article lend support to the assertion that there are multiple change trajectories from one regime of food consumption to another. Nevertheless, there are some similarities in the foodscapes of our diverse respondents, reflecting wider structural shifts in food habits. This allows us to conclude about increasing complexity of household foodscapes during post-socialist transition. For instance, the range of foods available has increased, along with the complexity of making choices among all the options now available. Food purchase habits have been affected with the arrival of supermarkets and the increased availability of foods, both common and foreign. Gardening and gathering of wild foods remain common practices in many households, and are strongly rooted in cultural traditions. Social networks in exchanging homegrown food with relatives, neighbors and friends remain relevant, albeit in lower importance, and food continues to nourish social ties. The socialist *blat* system has faded while charitable food procurement has appeared in response to poverty and changes in the market system and practices of self-provisioning, and collective purchasing and queuing are now memories of the past.

After providing interpretive foodscape vignettes of four diverse social worlds, we have reflected upon what the exploration of household foodscapes of the past and present might lead us to conclude about the possible broader patterns of change and shifts of household foodscapes in association with broader political, cultural and societal changes. These households have narrated the kinds of changes they have seen in the various acts of food consumption within their foodscape over the last two decades. Household food consumption practices echo macro societal changes while simultaneously demonstrate consumer capacity to construct, reflect, and adjust the meaningful food behaviors in their individual worlds.

Studies of food consumption practices can serve to illuminate how the emergence of new and changing political, economic, and social environments touches upon individual foodscapes and how new forms and opportunities of consumption influence existing food habits. We suggest that there is further work to be done with using everyday food consumption practices in the current context of ex-Soviet bloc countries. An exploration of differences of meanings and motivations of consumption practices across generations and across critical moments along one's life trajectory (e.g., having children, falling ill) could provide additional details lost in "snapshot" explorations. Similarly, deeper exploration into the differences and similarities along socioeconomic lines would likely add to a better understanding of the impacts of processes of social differentiation. As Eglitis (2011) argued, there is relatively little discourse on class, power, and inequality on consumption in post-socialist Latvia. Another possible avenue for further research is to take a comparative approach with foodscape narratives within the post-socialist region and their similarities or differences from other European countries.

Acknowledgments

We wish to express our thanks to all participants who gave their time generously. This article has benefited from the helpful comments from conference participants, where earlier versions of this article were presented, and from anonymous reviewers. We would like to express special thanks to Pēteris Ručevskis for his assistance with visual design of the household foodscape maps.

Disclosure Statement

No potential conflict of interest was reported by the authors.

Funding

This study is a part of Marie Curie PUREFOOD project "Urban, peri-urban and regional food dynamics: toward an integrated and territorial approach to food," EC Grant PEOPLE-2010-INT, 264719, and European Regional Development Fund project "Visualisation tools of dynamic networks", Grant [2010/0318/2DP/2.1.1.1.0/10/APIA/VIAA/104]. This article does not reflect the opinion of the European Commission.

Notes

1. We use the term "post-socialist" to define the current status of Latvia and other ex-Soviet bloc countries with caution. Referencing the current period as a function of its relation to its recent past can be problematic and may not adequately acknowledge the deep and varied transformations in economic and social life that former socialist countries have undergone in the last two decades. Furthermore, Olga Shevchenko has made a compelling case for the rejection of this term to describe the current context by arguing that "old socialist traditions, both narrative and pragmatic, were selectively revamped and infused with new life, and new social forms emerged as by-products of people's efforts to preserve stability" (Shevchenko 2009, 12). It is, however, beyond the scope of this article to engage with the debates of the appropriateness or usefulness of this term today.
2. Our thanks to an anonymous reviewer who called this important point to our attention. For a detailed summary of the discourses around social memory and post-Soviet change (see Rivkin-Fish 2009).
3. Concepts and objects which were important in the past are represented by a dark circular node situated closer to the thematic square node. Those which are important in the present are represented by a light circular node placed further from the thematic square node. In some instances there is no circle, only the label of a concept or object, and in this case the item was described as relevant in the household narrative by virtue of its absence in both the past and present.
4. The lighter colored circles represent actants which have increased in importance in today's foodscape compared to that of the past whereas the dark gray circles are actants which have decreased in importance. Their degree of change is on a bimodal vertical axis where actants which fall in the central section experienced little or no change (e.g. wild foods), and actants which fall toward the top and bottom of the image experienced a great deal of change (e.g. queuing and availability). Actants sitting toward the far left are not present at all in today's foodscape, whereas those sitting to the far right are very important in today's foodscape. Lastly, the size of the circles represents the number of households mentioning each actant and only those actants which were mentioned by at least two households are included.

References

Alanen, I., eds. 2004. "The Transformation of Agricultural Systems in the Baltic Countries – A Critique of the World Bank's Concept." In *Mapping Rural Problematics in the Baltic*, 5–58. Aldershot: Ashgate.

Alber, J., and U. Kohler. 2008. "Informal Food Production in the Enlarged European Union." *Social Indicators Research* 89 (1): 113–127. doi:10.1007/s11205-007-9224-1.

Banyte, J., L. Brazioniene, and A. Gadeikiene. 2010. "Expression of Green Marketing Developing the Conception of Corporate Social Responsibility." *Inzinerine Ekonomika-Engineering Economics* 21 (5): 550–560.

Bildtgård, T. 2009. "Mental Foodscapes: Where Swedes Would Go to Eat Well (And Places They Would Avoid)." *Food, Culture & Society* 12 (4): 498–523.

Bourdieu, P. 1984. *Distinction: A Social Critique of the Judgment of Taste*. London: Routledge.

Brembeck, H., and B. Johansson. 2010. "Foodscapes and Children's Bodies." *Culture Unbound* 2: 797–818.

Buchler, S., K. Smith, and G. Lawrence. 2010. "Food Risks, Old and New: Demographic Characteristics and Perceptions of Food Additives, Regulation and Contamination in Australia." *Journal of Sociology* 46 (4): 353–374. doi:10.1177/1440783310384449.

Caldwell, M. L. 2004. *Not by Bread Alone: Social Support in the New Russia*. Berkeley: University of California Press.

Caldwell, M. L. 2011. *Dacha Idylls: Living Organically in Russia's Countryside*. Berkeley: University of California Press.

Caldwell, M. L. 2002. "The Taste of Nationalism: Food Politics in Postsocialist Moscow." *Ethnos: Journal of Anthropology* 67 (3): 295–319. doi:10.1080/0014184022000031185.

Eglitis, D. 2011. "Class, Culture, and Consumption: Representations of Stratification in Post-communist Latvia." *Cultural Sociology* 5 (3): 423–446. doi:10.1177/1749975510379963.

Filomena, S., K. Scanlin, and K. B. Morland. 2013. "Brooklyn, New York Foodscape 2007-2011: A Five-year Analysis of Stability in Food Retail Environments." *International Journal of Behavioral Nutrition and Physical Activity* 10: 46. doi:10.1186/1479-5868-10-46.

Fischer, L. P. 2010. "Turkey Backbones and Chicken Gizzards: Women's Food Roles in Post-Socialist Hungary." *Food and Foodways* 18 (4): 233–260. doi:10.1080/07409710.2010.529018.

Glaser, B. G., and A. L. Strauss. 1967. *The Discovery of Grounded Theory: Strategies for Qualitative Research*. Chicago, IL: Aldine.

Goulding, C. 2002. *Grounded Theory: A Practical Guide for Management, Business and Market Researchers*. London: Sage.

Hansen, M. W., and N. H. Kristensen. 2013. "The Institutional Foodscapes as a Sensemaking Approach towards School Food." In *Making Sense of Consumption: Selections from the 2nd Nordic Conference on Consumer Research 2012*, edited by L. Hansson, U. Holmberg, and H. Brembeck, 299–312. Göteborg: University of Gothenburg.

Hervouet, R. 2003. "Dachas and Vegetable Gardens in Belarus Economic and Subjective Stakes of an Ordinary Passion." *Anthropology of East Europe Review* 21 (1): 159–168.

Horská, E., J. Ürgeová, and P. Renata. 2011. "Consumers' Food Choice and Quality Perception: Comparative Analysis of Selected Central European Countries." *Agricultural Economics* 2011 (10): 493–499.

Jehlička, P., T. Kostelecký, and J. Smith. 2013. "Food Self-provisioning in Czechia: Beyond Coping Strategy of the Poor: A Response to Alber and Kohler's 'Informal Food Production in the Enlarged European Union' (2008)." *Social Indicators Research* 111 (1): 219–234. doi:10.1007/s11205-012-0001-4.

Jehlička, P., and J. Smith. 2011. "An Unsustainable State: Contrasting Food Practices and State Policies in the Czech Republic." *Geoforum* 42: 362–372. doi:10.1016/j.geoforum.2011.01.005.

Klumbytė, N. 2010. "The Soviet Sausage Renaissance." *American Anthropologist* 112 (1): 22–37. doi:10.1111/j.1548-1433.2009.01194.x.

Krebs, V. 2000. "Working in the Connected World Book Network." *International Association for Human Resource Information Management Journal* 4 (1): 87–90.

Latour, B. 2005. *Reassembling the Social: An Introduction to Actor-Network Theory*. Oxford: Oxford University Press.

Law, J., and J. Hassard. 1999. *Actor Network Theory and After*. Oxford: Blackwell.

Ledeneva, A. 2009. "From Russia With Blat: Can Informal Networks Help Modernize Russia?" *Social Research* 76 (1): 257–288.

Miewald, C., and E. McCann. 2014. "Foodscapes and the Geographies of Poverty: Sustenance, Strategy, and Politics in an Urban Neighborhood." *Antipode* 46 (2): 537–556. doi:10.1111/anti.v46.2.

Mikkelsen, B. E. 2011. "Images of Foodscapes: Introduction to Foodscape Studies and Their Application in the Study of Healthy Eating Out-of-home Environments." *Perspectives in Public Health* 131 (5): 209–216. doi:10.1177/1757913911415150.

Mincytė, D. 2014. "Raw Milk, Risk Politics, and Moral Economies in the New Europe." In *Ethical Food Movements in a Comparative Perspective*, edited by M. A. Caldwell, Y. Jung, and J. B. Klein, 299–312. Berkeley: University of California Press.

Morgan, K. 2010. "Local and Green, Global and Fair: The Ethical Foodscape and the Politics of Care." *Environment and Planning A* 42: 1852–1867. doi:10.1068/a42364.

Patico, J. 2002. "Chocolate and Cognac: Gifts and the Recognition of Social Worlds in Post-Soviet Russia." *Ethnos: Journal of Anthropology* 67 (3): 345–368. doi:10.1080/0014184022000031202.

Patico, J., and M. L. Caldwell. 2002. "Consumers Exiting Socialism: Ethnographic Perspectives on Daily Life in Post-Communist Europe." *Ethnos: Journal of Anthropology* 67 (3): 285–294. doi:10.1080/0014184022000031176.

Ries, N. 2009. "Potato Ontology: Surviving Postsocialism in Russia." *Cultural Anthropology* 24 (2): 181–212. doi:10.1111/cuan.2009.24.issue-2.

Rivkin-Fish, M. 2009. "Tracing Landscapes of the Past in Class Subjectivity: Practices of Memory and Distinction in Marketizing Russia." *American Ethnologist* 36 (1): 79–95. doi:10.1111/j.1548-1425.2008.01110.x.

Rose, R., and Y. Tikhomirov. 1993. "Who Grows Food in Russia and Eastern Europe?" *Post-Soviet Geography* 34 (2): 111–126.

Round, J., C. Williams, and P. Rodgers. 2010. "The Role of Domestic Food Production in Everyday Life in Post-Soviet Ukraine." *Annals of the Association of American Geographers* 100 (5): 1197–1211. doi:10.1080/00045608.2010.520214.

Schwartz, K. Z. S. 2007. "The Occupation of Beauty: Imagining Nature and Nation in Latvia." *East European Politics & Societies* 21 (2): 259–293. doi:10.1177/0888325407299781.

Seeth, H. T., S. Chachnov, A. Surinov, and J. Von Braun. 1998. "Russian Poverty: Muddling through Economic Transition with Garden Plots." *World Development* 26 (9): 1611–1624. doi:10.1016/S0305-750X(98)00083-7.

Shevchenko, O. 2002. "'In Case of Fire Emergency': Consumption, Security and the Meaning of Durables in a Transforming Society." *Journal of Consumer Culture* 2 (2): 147–170. doi:10.1177/146954050200200201.

Shevchenko, O. 2009. *Crisis and the Everyday in Postsocialist Moscow*. Bloomington: Indiana University Press.

Sīmane, M., ed. 2003. *Latvia: Human Development Report 2002/2003: Human Security*. Riga: United Nations Development Program.

Smith, A. 2002. "Culture/Economy and Spaces of Economic Practice: Positioning Households in Post-Communism." *Transactions of the Institute of British Geographers* 27 (2): 232–250. doi:10.1111/1475-5661.00051.

Smith, J. 2003. "From Házi to Hyper Market: Discourses on Time, Money, and Food in Hungary." *Anthropology of East Europe Review* 21 (1): 179–188.

Smith, J., and P. Jehlička. 2007. "Stories around Food, Politics and Change in Poland and the Czech Republic." *Transactions of the Institute of British Geographers* 32 (3): 395–410. doi:10.1111/tran.2007.32.issue-3.

Šūmane, S. 2010. "Rural Innovation: Formation of New Development Practices. The Case of Biological Agriculture." PhD diss., University of Latvia, Riga.

Tisenkopfs, T., I. Kovách, M. Lošťák, and Š. Sandra. 2011. "Rebuilding and Failing Collectivity: Specific Challenges for Collective Farmers Marketing Initiatives in Post-Socialist Countries." *International Journal of Sociology of Agriculture and Food* 18 (1): 70–88.

Wenzer, J. 2010. *Eating Out Practices among Swedish Youth: Gothenburg Area Foodscapes. Cfk-Rapport 2010:3.* Goteborg: Goteborg Universitet.

Yamin-Pasternak, S. 2008. "A Means of Survival, a Marker of Feasts: Mushrooms in the Russian Far East." *Ethnology* 47 (2/3): 95–107.

Zagata, L. 2012. "Consumers' Beliefs and Behavioural Intentions Towards Organic Food. Evidence from the Czech Republic." *Appetite* 59 (1): 81–89. doi:10.1016/j.appet.2012.03.023.

Zavisca, J. 2003. "Contesting Capitalism at the Post-Soviet Dacha: The Meaning of Food Cultivation for Urban Russians." *Slavic Review* 62 (4): 786–810. doi:10.2307/3185655.

THE MAKING OF THE CONSUMER? RISK AND CONSUMPTION IN EUROPEANIZED LITHUANIA

Ida Harboe Knudsen

This article explores how larger geo-political changes are mirrored in the consumption habits in both rural and urban Lithuania. While the dominating European Union discourse emphasizes health, safety, and hygiene in food production, a counter-discourse has emerged where "authenticity" and "tradition" are embraced in the re-evocation of Lithuanian farm products. Based on ethnographic material, I suggest that the coexisting ambivalent and often contradicting feelings toward the West in present-day Lithuania have resulted in both a desire for and refusal of western products, as well as in a revival of the local cuisine.

My Encounter with a Refrigerator in Vilnius

In 2007, while conducting fieldwork in the Lithuanian countryside, I visited the capital Vilnius for a few days, where I stayed with my acquaintance Arnas and his wife.[1] Arnas was living in a newly renovated flat, filled with expensive furniture, a brand new TV set and plenty of other technical equipment. The flat was an accurate reflection of Arnas: he had a university degree, always dressed in fine suits for work, drove a fancy car, liked to dine out, stayed on top of the most recent developments, and frequently traveled abroad. In other words, Arnas was an image of the *new Lithuanian* as described by anthropologist Victor de Munck: the modern and cosmopolitan citizen of Lithuania who is oriented toward the "West" (2008).[2] Arnas was conscious about displaying this appearance in his everyday life. I was therefore rather surprised to open his refrigerator. Contrary to my expectations, it *was not* filled with exotic delicacies from fancy stores further reflecting his international image; it was stuffed with typical Lithuanian farm products such as home-made white cheese, glass bottles of creamy milk, big pieces of unpacked pork and dirty, and brown countryside eggs. Arnas must have noticed my somewhat surprised expression, as I faced a range of products identical to the ones I just had left behind me in the countryside. He smiled at me and explained that he and his wife preferred to get food products from the marketplace or the countryside, as "these were much healthier" than what you could get in the store. The latter were usually "filled with food preservatives" and "damaging for your health."

It was not the first time I heard this opinion; what Arnas had just said was everyday knowledge in the villages where I conducted research. Homegrown food, it was argued, was *real food*; incomparable with things you could obtain in the store. What surprised me in the case of Arnas was not this argumentation in itself, but the fact that these ideas were not only promoted and circulated by rural producers and consumers, but were also articulated by a man who otherwise made a deed out of being western. Driven by curiosity, I now started to take sneak peeks at the content of the refrigerators of other friends and acquaintances in the capital. I frequently encountered the same situation as with Arnas: typical Lithuanian countryside products found their way to consumers in the big city. Furthermore, as my inquiries revealed, these were often products which *had not* been European Union (EU) certified, and thus did not conform to the EU requirements for food safety and production.

In her study of raw milk production in Lithuania, Diana Mincyte noted that despite the introduction of EU standards and the zero-tolerance policy toward uncertified dairy products by the Lithuanian Health Agency, the raw milk market in Lithuania has not been weakened since the country's EU entrance in 2004. Many Lithuanians prefer local dairy products from the market and big foreign or Lithuanian companies have not managed to convince them otherwise, despite the growing awareness about food safety regulations (Mincyte 2009). What intrigued me was that the Lithuanians I knew from my fieldwork emphasized the local products' nutritional value as the main reason for buying them, despite the fact that they were not overly concerned with their health and diet in general. Arnas, for example, smoked a pack or more of Marlboro Lights a day, and gladly drank large amounts of beer. Still, the narrative of healthy Lithuanian farm products *being in opposition to* unhealthy imported products had come to dominate people's beliefs, and this appeared to be the case in both the rural areas, where I conducted my fieldwork, and among urban friends and acquaintances. Thus, ideas about what was perceived as natural food from rural producers had captured the imagination of a wide spectrum of the population, bearing witness to general concerns about changes and risk in society. I argue that the increased desire to obtain farm products is not an influence from western Europe, where anti-McDonaldization and eco-trends have been on the rise. Rather, the reinvention of authentic cuisine is a way of negotiating ambivalent feelings *toward* the West through the evocation of a distinct Lithuanian cuisine. Contrary to the western focus on low fat products as signifiers for health, the Lithuanian cuisine is characterized by the view on high-fat products as healthy and nutritious. Thus, the trend of reinventing the local cuisine is to be analyzed as a response to the West.

Ambivalence and Failed Person-Making

Based on my fieldwork among small-scale rural producers in the Lithuanian countryside, as well as among their rural and urban customers, I will explore how risk and fear are connected with the consumption of uncertified dairy products. Relying on anthropologist Frances Pine's argument that consumption is an expression of approval or rejection of current politics and ideology (2001), I suggest that the coexisting ambivalent and often contradicting feelings toward the West in present-day Lithuania have resulted in both a desire for and refusal of western products. On the one hand, we are witnessing a will to become "European," as expressed through an increased orientation toward the West in terms of the consumption of technology, clothes, cars, movies, and music, not to mention a general mobility Westwards (De Munck 2007, 2008; Vonderau 2007, 2010). On the other hand, we find new fears about diseases and infertility, along with fear about moral and cultural degradation as unavoidable aspects of the ongoing westernization of society, resulting from the import of both foreign food products and foreign ideas (Harboe Knudsen 2012).

I connect this preference for homemade products with what I coin as "failed person-making" (Harboe Knudsen 2010). Here, I build on Elizabeth Dunn (2004) who, in her analysis of the consequences of the EU standardization policies in Poland, argues that the implementation of standards is connected to a *person-making*. In order to organize production in the same way as in western Europe, Polish farmers must be transformed into obedient subjects who follow the system of EU rules and internalize the values and thereby become self-disciplined citizens through their working process (Dunn 2005, 2004). While farmers who have the ability, knowledge, and resources are being made into EU producers, the vast majority of small-scale farmers are failing this person-making. They are unable to buy new equipment and improve conditions on their farms sufficiently so as to participate in the EU standard regime. Unable to follow the rules and thereby unable to produce legally within the EU framework, they still cannot afford to lose the income from their sale of farm products. My ethnographic data suggest that instead of ending up as marginalized and excluded, farmers who produce outside of the standard regime draw on the current EU discourses promoting artisanal production by positioning their products as traditional and authentic, in opposition to manufactured products from the EU (Roberts 2007; Caldwell 2009; Klumbytė 2009, 2011; Mincyte 2009, 2014). The dominating EU discourse that emphasizes health, safety, and hygiene in food production is thereby met locally with counterarguments based on *tradition* in the re-evocation of Lithuanian farm products. Rather than embracing EU values, farmers thus advocate a different set of values attached to their production. This is what I refer to as a "failed person-making." Furthermore, as long as the consumers do not integrate EU-promoted ideas into their consumption habits, but rather favor the values of what they define as nutritious farm products, the illegal market for raw milk products has a high likelihood of survival. I analyze such processes by employing the concept of *EUropeanization*, the process of the implementation of the EU reforms in terms of the various, and often unintended outcomes that occur in local contexts (Harboe Knudsen 2012, p. 3–5).

Research and Field Sites

The findings of the research are based on one and a half years of participant observation in two regions of rural Lithuania mainly among small-scale farmers working on one to ten hectares of land. My first stay was for half a year in 2004, during which I lived and worked together with people from a village in rural Lithuania, which I here call Straigiai. This was followed by 1 year of research from 2006 to 2007. Here, I also included another village in the research project, here called Bilvytis, located in the neighboring region. During the two periods of research I conducted around 70 interviews with farmers, politicians, agricultural advisors, and other employees from the municipal administrations. The main part of the material was gathered through participant observation. My approach was to take part in the daily work on the farms and thus gain access to the farmers' daily routines and talks. Further participant observation was conducted at the marketplace, where women from the villages sold their farming products. I conducted informal interviews with people buying countryside products, in the village or at the marketplace, just as I paid attention to the distribution of farm products among relatives and friends. Finally, I paid frequent trips to the bigger cities to meet with agricultural experts. By both following discussions at the EU level, and witnessing the perception and implementation of the EU rules on small-scale farms, I was able to follow what Tomas Sikor (2005) refers to as the discrepancy between "legal rights" and "rights-in-practice." Hereby, it is meant that the official law may be changed in everyday life, as people do not follow the rules blindly but often react to the different restraints either by ignoring or directly opposing these rules.

Influences from the EU

The EU is not a recent phenomenon in the accession countries, but entered the stage and introduced the earliest preparation for obtaining the membership at the same time as the previously socialist countries struggled to undo the dominion over their societies. Starting from the early 1990s, the EU required that the candidate countries implement legislation from the *Acquis Communautaire*, the entire body of the EU law (Nello 2002). Farming was an important part of this, as the applicant states were largely dependent on their agricultural sectors (Fernandèz 2002). The EU together with the World Bank, the International Monetary Fund (IMF), and the Organization for Economic Co-operation and Development (OECD) played a central role in supervising the further development of the agricultural sectors in the previously socialist countries (Spoor and Visser 2001; Alanen 2004; Gorton, Lowe, and Zellei 2005). All of these organizations pushed for the creation of privately owned enterprises, as it was believed that these would be more efficient than collectively owned farms. Developments thereby rested on the idea that family farms could revive the agricultural sectors (Abrahams 1996; Alanen, Nikula, and Põder 2001; Alanen 2004). Throughout the 1990s, it became evident that rather than modernizing the agricultural sectors, the result of privatization had been a return to peasant forms of agriculture (Cartwright 2001, 2003). Furthermore, old Soviet power structures influenced the privatization process, as people with more influence and better connections also managed to get more land and property than the average worker (Verdery 1996). In Lithuania, this resulted in an agricultural sector consisting of more than 300,000 unspecialized small-scale farms based on semi-subsistence production, with a lack of resources, lack of equipment, and lack of know-how about the EU market (Harboe Knudsen 2012).

In order to integrate the agricultural production of the new member states into the overall production of the EU, the new states were required to meet specific standards for food production and food processing as detailed in the *Acquis Communautaire*. This meant that products from the new member states were expected to be of the same quality as those of their western European counterparts (Dunn 2005). The EU requirements are part of the central process of *harmonization* of products within the EU, which implies that products and their manufacture are to be governed by the same legislation in all member countries, according to a "one-size-fits-all" model. The problem arose in the new member countries that the methods of agricultural production on small-scale farms were from the outset too different to be encapsulated in the EU regime of standards and quality, and there was little or no chance of making them compatible as people lacked the necessary resources and knowledge. The result was that, when the EU laws were imposed on small-scale production, much of what hitherto had been occurring inside the boundaries of the law was now classified as illegal. Furthermore, they were now seen as potentially dangerous, as there had not been an EU-supervised control of their production, just as no bacteriological tests had been carried out on the final products (Dunn 2005; Caldwell 2009; Harboe Knudsen 2010).

EUropeanization and Perceptions of Risk

In the beginning of the article, I presented two central points. I first suggested that the consumers' preference for countryside products could be interpreted as a way of expressing their ambiguous feelings toward increasing modernization and globalization. Second, I identified the new discursive promotion of Lithuanian farming products that frames local production as authentic and as being in opposition to the synthetic food from the West. I refer to such negotiations and implementations based on the local environment as *EUropeanization* – not to be mistaken with *Europeanization* (Harboe Knudsen 2012). The latter commonly confuses the complex geographical constellation of Europe with the EU, and in this way interprets Europeanization as an outgoing movement where (old) Europe exercises its influence over socialist countries (Ågh 1993; Grabbe 2005; Hugh, Sasse, and Gordon 2005; Bafoil 2009). Thus, by uncritically using the concept of Europeanization, we run the risk of reproducing ideas of the EU as representing both Europe and a certain correct form of European-ness that is superior to various other ways of being European.

By using the concept *EUropeanization*, I refer to the specific geopolitical entity of the EU (thus not confusing it with Europe as such), just as I stray away from the very colonial perception of power influence from "West" to "East." I define the concept in terms of the socioeconomic and legal outcomes and changes in self-perception targeting the population at large and resulting from the coexistence of old Soviet structures and goal-oriented EU influences. The illegal dairy food markets and the consumers' preference for real countryside products as contradicting the goals of industrialization of the EU's agricultural sector demonstrate that local negotiations of rules and ideas lead to unintended outcomes. To better understand the context in which such unintended outcomes emerge, it is pertinent to locate informal economies in relation to current trends of consumption and the shortcomings of the EU's agricultural standards regime.

Ever since the breakup of the Soviet Union, radically new consumer trends have been noticeable in all previously socialist countries. Options for consumption changed from scarcity and homogeneity (Verdery 1996) to a wide range of goods and heterogeneity, visibly demonstrated through mushrooming of big shopping malls in the former Soviet countries. This made eastern European consumers capable of expressing the means of identity through the products they bought (Patico and Caldwell 2002). Consumerism was not a new phenomenon for eastern European citizens who had also used goods as ways to express their relations to each other and the state during the socialist period. However, with the breakdown of the Soviet Union, local consumer culture acquired different features as consumption options became more plentiful (Patico and Caldwell 2002). This is demonstrated not only in terms of what people consume, but also in terms of what they choose not to consume. Contrary to the often visibly displayed preferences for western goods, I have witnessed in terms of clothes, cars, and all sorts of technical equipment, changes seem to move at a far slower pace when it comes to consuming food products.

The consumers' preference for homemade food products is of vital importance for small-scale producers in Lithuania. As Mincyte argues, the people threatened most by increased EU control and regulations are Lithuania's *new poor* who are dependent on the illegal (and thus cheap) products in order to get by on low incomes and small pensions (2014). She thus proposes that the EU confuses two forms of risk when implementing new standards. While one of the primary goals in the EU's agricultural sector is to eliminate public health risks by enforcing food safety regulations, the participants in informal economies are focusing on economic risks, as their main priority is surviving through critical economic circumstances (2014). We cannot even adequately argue that the *new poor* indeed do pose a greater risk to their health by consuming uncertified products. As anthropological research has shown (Dunn 2005), measuring eastern European production on a scale emerging from the large industrial farms of western Europe has only shown the ridiculousness of the EU regulations, as umbrella models cannot be adequately transferred to a new context without serious implications. In her study of Polish meatpacking industry, Dunn has shown how Poland was considered a high-risk country for Creutzfeldt-Jakob disease (mad cow disease). This conclusion was reached despite the country having never had a single case of it before being subjected to EU regulations – this being *due* to its low technological production, which did not include feeding cattle with industrial food based on processed bone meal from cows, the way it is done in many western European countries (Dunn 2005). Similarly, Zsuzsa Gille's research on Hungarian paprika has shown us how several scandals concerning toxic Hungarian paprika only emerged as a result of the country's engagement with the EU market, which featured *less* control and higher risk factors than what was experienced before the country's EU entrance (Gille 2009). Furthermore, introducing more regulations is in itself not a direct way to implement standards and secure their function. Research both in Poland and Lithuania has proven that EU policies have pushed many small-scale farmers out of legal production, making space for an increase in the uncontrolled black market (Dunn 2004, 2005; Harboe Knudsen 2010, 2012; Mincyte 2009, 2014). The above examples resonate with my ideas of EUropeanization as a disparity between models and actual outcomes.

If we return to Mincyte's (2014) proposition of economic risk as a stimulus to stay in the illegal market, this certainly has proven to be true for the marginalized poor in the present-day society. Yet, it does not solely explain why people like Arnas, who can afford to buy far more expensive products, still have a preference for the cheaper goods from the Lithuanian countryside. I suggest that Lithuanian consumers are concerned with yet an additional risk, which is connected to the perception of a too radical and too fast "westernization." Risk in this sense becomes a matter of perception and experience, rather than being tied to certain objective features in society (Giddens 1991). The emotional and personal aspects obtain a central position, as both insecurity experienced in the past and the expectation of profound changes in the future are likely to undermine people's general sense of security (von Benda-Beckmann and von Benda-Beckmann [1994] 2000). The anxiety about the day of tomorrow finds expression in the evocation of exogenous risks that threaten to undermine the foundations of a healthy Lithuanian society. This is particularly articulated through a range of negative images, spanning from the "fabrication" of animals raised on hormones, genetically modified organism (GMO) and dairy products containing preservatives, to a general fear of increased individualization within society, moral degradation, broken families, and immoral and promiscuous behavior. Today, we witness how such scary images, intertwined with the celebration of materialistic culture, are circulating in a range of former socialist countries (Pine 2001; Caldwell 2009).

Buying at the Market

The marketplace in the city of Marijampolė provides a good example of such responses to the EU-standard regime. The market consists of a hall, where dairy and meat products are sold. The market authorities strictly control this space. The salespeople inside have had their products thoroughly checked and have obtained a special certificate for sale, which proves that their products are produced according to EU regulations. During my fieldwork, I went there with a woman in her early 30 s named Jovita, the daughter of the previous agronomist at the former collective farm, and an older woman from one of the wealthier families in the village, named Stasė. The people who sell in the market hall mostly come from better-off families with a higher level of education than the average villagers. The women from my field site have specialized in dairy products. The most popular product is *grietinė*, a product similar to sour cream. The Lithuanian white, compressed, egg-shaped cheese (*baltas sūris*) is also a valued product, as is *varškė*, curd packed in small plastic bags. A few of the women also sell unprocessed milk. All products are kept in open refrigerators, ensuring that they are kept at the right temperature.

Outside the official market territory, another space for selling has formed. Here, on the boundary of the market, women from the nearby countryside offer their dairy products. Here, we find baskets with white cheese and *varškė*, big milk bottles, and buckets of *grietinė*, the very same products that are sold inside. While there are 6 women selling inside, the number of illegal salespeople outside varies from 6 to 13. The women selling outside do not have the resources, money or knowledge to produce in accordance with the EU regulations. As they have not invested in special equipment for production and because they do not need to pay extra fees for tests and certificates as required in the market, they can sell their products at a lower price. Their physical location is cleverly chosen. Being just outside the market boundary, the authorities of the marketplace cannot claim control over their products or fees for their sale. Nevertheless, every customer who wishes to enter the market from the main street will pass the row of illegal salespeople first, which gives everybody an opportunity to buy their dairy products here at a cheaper price. In this way, the market is literally being divided into two spaces; a "legal inside" and an "illegal outside."

This division of the market is not a direct result of the EU-standard regime. Before the EU entrance in May 2004, there had also been rules for selling at the marketplace and not all producers were capable of fulfilling them or paying the money for selling legally inside. However, after the EU entrance the number of illegal salespeople increased. Many of the women, who had previously been selling inside the market hall, next to other certificated farmers, lost their right to sell their milk and cheese, as they could not meet the EU regulations. Because they could not afford to lose the income from the sale of their dairy products, they, rather than withdrawing from commercial production as demanded, instead started selling illegally outside the bounds of the market.[3] Although the products did not change, they were now evaluated according to a different scale of quality and production, namely the EU standard regime, where they no longer could fulfill the criteria. The immediate result was thus that rather than enforcing legal sale through strict regulations, the number of people selling illegally grew as a direct result of the introduction of the new standards and regulations.

Defining a Healthy Product

The elevation of national products in contrast to the food safety policy backed by the EU resulted in conflicting opinions in daily life about what constituted a safe and healthy food product. What appeared confusing for many consumers was that products, which hitherto had been known as healthy, now were banned by the EU. A man who went to buy at the market in Marijampolė explained to me that he had been eating products straight from the farm since he was a child. Now, all of a sudden, these products were potentially dangerous to his health. He did not take this at face value:

> I have never had any health problems. So, if these [farming] products really were bad, I would be sick or dead by now. But no, I am strong and healthy. I grew up in the countryside. I always ate homemade cheese and sausages. That was a long time before we knew about such as things as the EU. [Y]ou just look to western Europe. Are people healthy there? My son works in England and tells me how fat the British people are – munching bad food in front of the TV. Are Lithuanians supposed to become like them? Is this the goal? (Harboe Knudsen, Interview, May 2007).

The problem for Lithuanian health, food, and agricultural development institutions was the persisting discontinuity between the official and scientific discourses and everyday conceptualizations of what a good food product was. New understandings of hygiene often conflicted with perceptions of healthy milk coming straight from the cow. These understandings were equally articulated among illegal producers from the countryside and the customers who purchased their products.

The EU regulations imposed strict regulations on dairy production. These included the requirement that milk should not contain residential antibiotics, just as the products were required to be regularly checked for bacterial contamination. Dairy products could only be produced in properly equipped rooms fitted with a certain kind of tile, and the sterilization of hands and equipment was also required. Most of the informants from the villages where I conducted my research were engaged in illegal sale, Jovita and Stasė being the only exceptions. Others produced dairy products with the means they had at hand, using the rooms for storage that were available to them. Like the women selling outside the bounds of the market, the uncertified producers in the villages did not lack customers. They all had well-developed networks of neighbors, relatives, friends, acquaintances and more distantly connected consumers who flocked to the villages whenever a pig was slaughtered, or when fresh cheese was ready for sale. They shared a common understanding that the products made at home were far healthier and tastier than cheeses and sausages available in the store. Fat and the taste of fat played a vital role in the perception of high-quality and traditional Lithuanian food. At the marketplace, for example, a product with more fat meant a better product; a way to show the quality of the *grietinė* was to insert a long spoon into the bucket. If the spoon could stand up straight, it meant that the *grietinė* was high in fat and thus of high quality. Furthermore, if cheese is not fatty, the women told me, it does not have any substance, and you have to sell it for less money. They laughed and shook their heads when I told them that in my country people often pay more for low-fat dairy products, because they want to stay slim.

The natural-food wave I thus experienced in Lithuania was born out of a comparison with the new types of production within the EU. It was mainly directed toward imported products, which were representing ideas of fake nutrition, but it was likewise directed toward similar Lithuanian products, which had passed through the range of tests required by the EU. Rather than viewing certified sellers from the countryside as being superior due to their fulfillment of standards, their products were now discursively connected to the artificial food imported from the EU nutrition and food preservatives. Indeed, if the EU now accepted their products, they could no longer be traditional in the same way – this was the argumentation among uncertified sellers, as they discussed such issues during extended coffee-drinking sessions. It was also suggested that the legal sellers most likely had started to use various chemicals in their production in order to get their cheese to look better and last longer.

The suspicion raised among the consumers regarding the products approved by the EU was based on attempts to rearticulate the value of "traditional" food. In an analysis of two competing French biscuit companies, Simon Roberts (2007) shows how a regional product becomes a product of heritage and tradition only by the commercialization of purposefully created history. Roberts is interested in the moment when the unspoken and previously taken-for-granted idea of a given product is made explicit and consciously articulated, namely, when a product gains its originality and heritage by becoming a part of a commercial discourse. Its authenticity as a traditional product becomes a prerequisite for its quality, and the present regains new meaning in the light of an idealized past (Roberts 2007). If we relate these thoughts to homemade food products from Lithuania, we also follow a pattern by way of which they change from being merely food and become consciously traditional Lithuanian products. In this way, they become connected to a more complex understanding that goes far beyond cheese: the negative image of capitalism, and a reflection of ambivalence toward the EU, which is seen as possibly representing a way to modernity, but possibly as being a threat to society due to its "fake" values.

Consumption: Expressions of Rejection and Embracement

Some years ago, I read an interview in an American popular magazine (*Men's Health*) with Walter Scheib, the former chef for President Bill Clinton and later for President George W. Bush. In the interview, which was given in light of Scheib's book release *White House Chef: Eleven Years, Two Presidents, One Kitchen* (2007), he explained how the eating habits in the White House had changed after 9/11. During this period, Scheib remarked, President Bush and his family requested more traditional American food, food that resembled a feeling of home, while they avoided foreign and what they considered as exotic food. The point may be farfetched, but by reading the article I drew certain parallels between a situation of crisis, as experienced in the USA, and the general feeling of insecurity among many Lithuanians in light of the rapid and massive restructurings of society after the Soviet breakdown. Both examples give voice to a preference for foodstuffs that symbolize home and generate feelings of safety. Imported and exotic food was not capable of generating the same feelings. Indeed, Lithuanians from my field site cultivated a distinct culture about their own cuisine, which not only had a specific taste, but was also characterized by claims to having a higher quality of food than products from western Europe.

If we compare with anthropologist Neringa Klumbytė's (2009) article about the consumption of sausages in Lithuania, Soviet products are still highly popular in Lithuania. Klumbytė examines consumers' opinions about and consumption of two different sausage brands in Lithuania: one produced by the Samsonas company and called "Tarybinė" (Soviet) for its resemblance of the supposedly traditional Soviet sausage and the other marketed by the Biovela company; the "Euro-Sausage." The preference of the consumers was clear: Samsonas' profit skyrocketed after it began to produce "Soviet" sausages [. . .]. In 2005, all "Soviet" meat brands comprised more than 50% of Samsonas' production (Klumbytė 2009, 130). According to Klumbytė, the Soviet sausage was perceived by her informants as familiar and part of the traditional Lithuanian cuisine. In addition, the Soviet sausage was recognized as a "natural" product without food preservatives. In comparison, the distinct "Euro-sausage" – which fulfilled EU criteria – never gained the same kind of popularity, and was never characterized as being part of the Lithuanian traditional cuisine (Klumbytė 2009). Klumbytė views this reestablishment of an imagined Soviet consumer identity as a response to the relative political dislocation experienced by the new member states; they perceive themselves as being "Europe's province" due to the unequal distribution of wealth and power in the EU (2011).

During my fieldwork, I likewise discovered the pride and sentiment attached to food from "home." On numerous occasions, I have been told how no honey is sweeter than the Lithuanian, no meat is tastier than the Lithuanian, no milk is healthier, no cheese is richer, and no vodka has the quality and taste the Lithuanian home-distilled has. Imported products from the West were not only foreign to the Lithuanian sense of taste; they also stimulated a sense of danger for diseases and decreasing health. Rather than being concerned with conditions for production at non-EU certified farms, the people I encountered in my research were more concerned with the mysterious food preservatives and chemical additives present in EU-certified products.

One informant, a 46-year-old man who was planning to start up his own organic farm, gave a telling example of the dangers imposed on Lithuanian society from the West. He narrated how he had come across the fact that a new "species" of people had emerged as a result of GMO products in the USA, changing their genetic code so that they now were born with severe deformations: long strips of skin hanging down all over their bodies, bearing a resemblance to boiled spaghetti. Clearly, as he further explained, this was kept hidden and the people in question were persuaded to live in disguise in order not to alarm the general public. He went on to argue that if products like these were to be introduced in Lithuania–and according to him this was an unavoidable consequence of opening the door to the West–this would not only lead to the same severe consequences as in the USA, but could also possibly cause infertility, and thus a decline in reproduction could be the consequence for the coming generations.

Stories like this bear witness to how various forms of consumption inevitably adopt political connotations as an approval, or rejection, of geopolitical changes and ideologies. More than just a question about economic risks, and more than emerging solely from a deep concern about one's health (while telling this story the man was consuming an impressive amount of beer and smoking one cigarette after another), the consumption of certain products is entwined with a general fear for the unknown dangers from the West. In the worst-case scenario these were capable of changing the very human being into a grotesque creature (with "spaghetti skin") and putting an end to the reproduction of mankind.

Frances Pine (2001), in an article about Poland shows how the attitude toward goods imported from the West has changed during the past decades. From being the status symbol of western culture during the socialist regime and being, thus, in opposition to it, the disappointments following the breakup of the Soviet Union made western products the very symbol of betrayed hopes and promises. The result was a turnaround in consumption practices, which took the form of the revival of national Polish products. In other words, Polish material culture and food products experienced a comeback as a way of voicing disappointment with the West. With regard to my fieldwork in Lithuania I found many mechanisms similar to what Pine's study shows us. Yet, the political, economic and social ups and downs had not resulted in an absolute rejection of western values by Lithuanians. Many people were still fascinated by the West, had hope for increased wealth in the future, and aimed at adopting a modern and western lifestyle (De Munck 2008). However, coexisting with expectations or hopes for increased wealth was a fear of moral degradation in Lithuania due to the quick modernization of industry and the subsequent proliferation of low-quality products (Caldwell 2009). Building on the argument proposed by Pine (2001) that consumption is an expression of approval or rejection of current politics and ideology, I suggest that Lithuanian consumers articulate a sense of "home" and "our-ness" through their preference for what now is viewed as traditional food products.

Concluding Remarks

My acquaintance Arnas from Vilnius represents two dominating tendencies in present day Lithuania: embracement and rejection. As I have argued in this article, such contradicting patterns of consumption mirror Lithuanians' often ambivalent attitude to the West generally and the EU specifically, as the concept of "abroad" is constantly negotiated both as a place of excitement and modernity and as a source of new risks. This, I have argued, is displayed particularly with regards to the consumption of uncertified food products, as the continued existence of illegal markets for farming products bears witness to very non-EU attitudes toward safety and risk. More than that, the idea of inherent danger has led to the emergence of new discourses of health, particularly when uncertified producers have reframed their products as "healthier" than imported products or even those produced legally in Lithuania.

If we return to the discussion about the EU's ambition to create new and self-controlled producers by the introduction of certain standards, as shown in Dunn's study (2004), the present production and consumption patterns undermine these very attempts. This is an example of what I coin as EUropeanization, as frameworks and regulations from the EU are negotiated locally, leading to other outcomes than originally were intended. Risk in the EUropeanized sense is thereby related to far more substantial threats than microbiological tests of grandmother's cheeses would reveal.

Disclosure statement

No potential conflict of interest was reported by the author.

Funding

Research was funded by EU and Marie Curie Programme for Social Anthropology.

Notes

1. Pseudonym. All other names of informants in this article are likewise pseudonyms.
2. When I use the concepts the "West" and "western" it is to be seen as a general reference to western Europe and the USA, and as a way to stay loyal to my informants' general references to the same regions.
3. As I carried out fieldwork at the marketplace in Marijampolė both during my first stay in Lithuania from January to July 2004 and again during my second stay from October 2006 to October 2007, I am able to draw on field data from both periods of research.

References

Abrahams, R. 1996. "Introduction: Some Thoughts on Recent Land Reforms in Eastern Europe." In *After Socialism: Land Reform and Social Change in Eastern Europe*, edited by R. Abrahams, 1–22. Oxford: Berghahn Books.

Ågh, Å. 1993. "Europeanization through Privatization and Pluralization in Hungary." *Journal of Public Policy* 13 (1): 1–35. doi:10.1017/S0143814X00000921.

Alanen, I., ed. 2004. *Mapping the Rural Problem in the Baltic Countryside: Transition Processes in Rural Areas of Estonia, Latvia and Lithuania*. Aldershot: Ashgate.

Alanen, I., J. Nikula, and H. Põder. 2001. *Decollectivisation, Destruction and Disillusionment: A Community Study in Southern Estonia*, edited by R. Ruotsoo. Aldershot: Ashgate.

Bafoil, F. 2009. *Central and Eastern Europe: Europeanization and Social Change*. New York: Palgrave Macmillan.

Benda-Beckmann, F. V., K. V. Benda-Beckmann, and F. V. Benda-Beckmann. (1994) 2000. "Coping with Insecurity." In *Coping with Insecurity: An 'Underall' Perspective on Social Security in the Third World*, edited by Benda-Beckmann, K. V., and H. Marks, 7–34. Yogyakarta: Pustaka Pelajar.

Caldwell, M. L. 2009. "Introduction: Food and Everyday Life after Socialism." In *Food and Everyday Life in the Postsocialist World*, edited by M. Caldwell, 1–28. Bloomington, IN: Indiana University Press.

Cartwright, A. 2001. *The Return of the Peasant: Land Reform in Post-communist Romania*. Aldershot: Ashgate.

Cartwright, A. 2003. "Private Farming in Romania: What Are the Old People Going to Do with Their Land?." In *The Postsocialist Agrarian Question: Property Relations and the Rural Condition*, edited by C. M. Hann, 171–188. Münster: Lit Verlag.

De Munck, V. 2007. "First, Second and Finally Third Order Understandings of Lithuanian National Identity: An Anthropological Approach." *Sociologija. Mintis Ir Veiksmas* 1 (19): 51–73.

De Munck, V. 2008. "Millenarian Dreams: The Objects and Subjects of Money in New Lithuania." In *Changing Economies and Changing Identities in Post-socialist Eastern Europe*, edited by I. Schröder, and A. Vonderau, 171–191. Berlin: Lit Verlag.

Dunn, E. C. 2004. *Privatizing Poland: Baby-food, Big Business and the Remaking of Labor*. Ithaca, NY: Cornell University Press.

Dunn, E. C. 2005. "Standards and Person-Making in East Central Europe." In *Global Assemblages: Technology, Politics and Ethics as Anthropological Problems*, edited by A. Ong, and S. J. Collier, 173–193. Malden, MA: Blackwell Publishing.

Fernandèz, J. 2002. "The Common Agricultural Policy and EU Enlargement: Implications for Agricultural Production in the Central and East European Countries." *Eastern European Economics* 40 (3): 28–50.

Giddens, A. 1991. *Modernity and Self-Identity: Self and Society in the Late Modern Age*. Cambridge: Polity.

Gille, Z. 2009. "The Tale of the Toxic Paprika. The Hungarian Taste of Euro-Globalization." In *Food and Everyday Life in the Postsocialist World*, edited by M. Caldwell, 57–78. Bloomington, IN: Indiana University Press.

Gorton, M., P. Lowe, and A. Zellei. 2005. "Pre-accession Europeanisation: The Strategic Realignment of the Environmental Policy Systems of Lithuania, Poland and Slovakia towards Agricultural Pollution in Preparation for EU Membership." *Sociologia Ruralis* 45: 202–223. doi:10.1111/soru.2005.45.issue-3.

Grabbe, H. 2005. *The EU's Transformative Power: Europeanization through Conditionality in Central and Eastern Europe*. New York: Palgrave Macmillan.

Harboe Knudsen, I. 2007. "Max Planck Institute for Social Anthropology." Interview, Pp. 3–4.

Harboe Knudsen, I. 2010. "The Insiders and The Outsiders: Standardization and 'Failed' Person-Making in a Lithuanian Market Place." *The Journal of Legal Pluralism and Unofficial Law* 42: 71–93. doi:10.1080/07329113.2010.10756650.

Harboe Knudsen, I. 2012. *New Lithuania in Old Hands: Effects and Outcomes of Europeanization in Rural Lithuania*. London: Anthem Press.

Hugh, J., G. Sasse, and C. Gordon. 2005. *Europeanization and Regionalization in the EU's Enlargement to Central and Eastern Europe: The Myth of Conditionality*. New York: Palgrave Macmillan.

Klumbytė, N. 2009. "The Geo-Politics of Taste: The 'Euro' and Soviet Sausage Industries in Lithuania." In *Food and Everyday Life in the Postsocialist World*, edited by M. Caldwell, 130–153. Bloomington, IN: Indiana University Press.

Klumbytė, N. 2011. "Europe and Its Fragments: Europeanization, Nationalism, and the Geopolitics of Provinciality in Lithuania." *Slavic Review* 70 (4): 844–872.

Mincyte, D. 2009. "Self-Made Women: Informal Dairy Markets in Europeanizing Lithuania." In *Food and Everyday Life in the Postsocialist World*, edited by M. Caldwell, 78–100. Bloomington, IN: Indiana University Press.

Mincyte, D. 2014. "Raw Milk, Risk Politics, and Moral Economies in the New Europe." In *Ethical Eating in the Postsocialist and Socialist World*, edited by M. L. Caldwell, Y. Jung, and J. Klein, 25–43. Berkeley, CA: University of California Press.

Nello, S. S. 2002. "Preparing for Enlargement in the European Union: The Tensions between Economic and Political Integration." *International Political Science Review* 23 (3): 291–317. doi:10.1177/0192512102023003005.

Patico, J., and M. Caldwell. 2002. "Consumers Exiting Socialism: Ethnographic Perspectives on Daily Life in Post-communist Europe." *Ethnos* 67 (3): 285–294. doi:10.1080/0014184022000031176.

Pine, F. 2001. ""From Production to Consumption in Postsocialism?." In *Poland Beyond Communism: 'Transition' in Critical Perspective*, edited by M. Buchowski, 209–224. Freiburg: Universitätsverlag Freiburg/Schweiz.

Roberts, S. 2007. "Order and the Evocation of Heritage: Representing Quality in French Biscuit Trade." In *Order and Disorder: Anthropological Perspectives*, edited by K. V. Benda-Beckmann, and F. Pirie, 16–33. Oxford: Berghahn Books.

Scheib, W., and A. Friedman. 2007. *White House Chef. Eleven Years, Two Presidents, One Kitchen*. Hoboken, NJ: John Wiley and Sons.

Sikor, T. 2005. "Property and Agri-Environmental Legislation Europe." *Sociologia Ruralis* 45: 187–201. doi:10.1111/soru.2005.45.issue-3.

Spoor, M., and O. Visser. 2001. "The State of Agrarian Reform in the Former Soviet Union." *Europe-Asia Studies* 53: 885–901. doi:10.1080/09668130120078540.

Verdery, K. 1996. *What Was Socialism and What Comes Next?*. Princeton, NJ: Princeton University Press.

Vonderau, A. 2007. "Yet Another Europe? Constructing and Representing Identities in Lithuania Two Years after EU Accession." In *Representations on the Margins of Europe, Politics and Identities in the Baltic and South Caucasian States*, edited by T. Darieva, and W. Kaschuba, 220–242. Frankfurt: Campus Verlag.

Vonderau, A. 2010. "Models of Success in the Free Market: Transformations of the Individual Self-Representation of the Lithuanian Economic Elite." In *Changing Economies and Changing Identities in Postsocialist Eastern Europe*, edited by I. Schröder, and A. Vonderau, 111–129. Münster: Lit Verlag.

ATLANTIC HERRING IN ESTONIA: IN THE TRANSVERSE WAVES OF INTERNATIONAL ECONOMY AND NATIONAL IDEOLOGY

Kadri Tüür and Karl Stern

This article examines the background of and reasons for a sudden decrease in the Estonian import of Atlantic herring during the Great Depression in 1932. The economic and ideological factors that influenced the process are discussed, including protectionist trade policy measures, customs regulations and nontariff trade measures. We argue that the attempt to replace herring imports by establishing a national herring fishing fleet was grounded in ideological as well as in nutritional arguments. Such protectionist measures were met with confrontation by Estonian foreign trade partners. The case study highlights a complicated interplay between oceanic resource exploitation politics and national ideologies, locating it in the context of regional environmental historical research.

As a renowned environmental historian, Helen M. Rozwadowski (2001, 221) has pointed out, oceans are not just an immensely rich environment and a source of livelihood, but they are also a contested playground of national powers and commercial interests. In this article, we combine textual analysis and research on economic history in order to present a case study of Estonia's attempt to develop a national fishing fleet in the northern part of the Atlantic Ocean in the 1930s, the period of the Great Depression and the subsequent years when European nation states, including Estonia, devised and tested different solutions for the unfavorable economic situation. The central object of our study is Atlantic herring as a food and trade object that occupied an important place in Estonia's food economy and that became the subject of the highly controversial policy, leading to the drastic drop in fish imports in 1932. We locate the Estonian case study in the context of European agricultural commodity markets, primarily in relation to the United Kingdom as one of the main trade partners of the Republic of Estonia of the time, as well as in relation to international herring fishing activities. Our purpose is to highlight the cultural and ideological constructions of the economic categories of domestic and foreign goods that were implemented in the Atlantic herring case. Such a categorization is an aspect of the foreign trade history that has not yet received attention in the earlier historiography.

It is generally acknowledged that when facing economic depression many countries sank into heavy protectionism in the 1930s. Liberal policies were both unpopular and uncommon, and many European countries, including Estonia, introduced a wide range of nontariff trade measures in efforts to protect their economies (Klesment 2000; Raud 1934/35). European as well as Estonian research on the economic history of this period mainly deals with the extent of and reasons for the protectionist policies in the general context. What has not yet been covered in Estonian research is the detailed analysis of designating a particular group of goods as domestic or foreign in the context of global market competition. Our research presented in this article addresses this issue.

In her monograph on Estonian foreign trade in 1918–1940, Estonian economic historian Maie Pihlamägi notes that the import of Atlantic herring that had comprised 90% of Estonian fish imports during the earlier years dropped drastically in 1932. She suggests that one of the reasons for such a decrease might be a change in the preferences of consumers, who opted for domestic meat and butter instead of the more expensive imported fish (Pihlamägi 2004). Foreign trade statistics seem to support this dynamic: Atlantic herring (and the relatively insignificant quantities of imported Baltic herring and pilchard) formed 1.3% of the whole import volume of the Republic of Estonia in 1931, but only 0.3% in 1932. In 1933, the share of Atlantic herring in Estonian imports rose again, remaining around 1% until 1939. Pihlamägi's argument that there was a relation between imported herring and domestic agricultural products provides a hypothesis to be studied, namely whether domestic products were set into opposition with foreign products and to what extent. Another and by no means less important question concerns the nature of the domestic products that were actually preferred over the imported Atlantic herring. The revision of the categories of domestic and foreign is pertinent here, as according to statistics, Atlantic herring was one of the key import articles in the trade group of food, spices, and drinks. In 1930, grain (both seed and ground), sugar, and herring were the top import articles in this trade group, forming respectively 13.2%, 5.2%, and 1.3% of Estonian import (Pihlamägi 2004). At the same time, local consumers already had an alternative to the consumption of Atlantic herring in the form of locally caught fish.

Our task in this project is to explore the possible reasons for the sudden but short-term drop in herring imports to Estonia in 1932. More specifically, we ask: were the forcefully implemented protectionist politics, combined with the propaganda in favor of domestically produced food items, among the major causes for the decrease in herring imports? The article provides a survey of the trade policy measures, especially the nontariff measures, implemented in Estonia during the Great Depression, that were used in the case of Atlantic herring.

Second, the ideological background for the "domestication" of the Atlantic herring is to be analyzed. This will be done on the basis of the media coverage of the "herring issue," especially in Estonian daily newspapers and in periodicals devoted to the advancement of the Estonian ocean fishing industry.[1] We will particularly focus on the travelogue of Estonian writer, essayist, and playwright Evald Tammlaan (1904–1945) who documented the first open-sea fishing trip organized by the Estonian company OÜ Kalandus [Fishery Company].

The third important question concerns the reasons why the sudden decrease in herring imports only proved to be short term. The reasons, as we may hypothesize, are again both economic and ideological. We ask whether the consumption of domestic food was propagated in Estonia in the 1930s at the expense of the imported Atlantic herring, and whether such protectionist activities met confrontation from Estonia's foreign trade partners.

The main sources related to the implementation of different trade policy measures in the 1930s are legislative acts and treaties of commerce, published in the state legislative periodical *Riigi Teataja* [State Gazette]. The reasons for their enforcement could be gleaned from archival documents belonging to the materials of several state institutions, such as the Ministry of Commerce, the State Chancellery, and the Parliament, held in the State Archives.[2] The primary source for the statistics of commerce is the periodical *Väliskaubandus* [Foreign Trade Bulletin], the official gazette for the publication of data on Estonian foreign trade.

Herring Imports and Protectionist Trade Policy

Salted Atlantic herring (*Clupea harengus harengus*) has formed a substantial part of European trade since the Middle Ages (Sicking and Abreu-Ferreira 2009; Põltsam 2008). Unlike many other seafood items, salted herring can be transported and stored over long distances and periods of time without losing its nutritional and gustative qualities. As its name indicates, Atlantic herring (*Clupea harengus harengus*) inhabits the waters of the Atlantic Ocean, and it usually grows to be 37 centimeters long, while in the coastal waters of Iceland, it can reach up to 42 centimeters. Individual specimens can weigh up to 0.5 kilograms. Herrings are oily fish whose meat is rich in Omega-3 fatty acids. It is also a source of vitamin D, having thus been an important addition to the predominantly grain-based menus of peasants.

The inhabitants of Estonia did not participate in catching or processing Atlantic herring. Nor did they encounter the species in the Baltic Sea as it prefers saltier waters. Herring reached Estonian consumers as a salted good ready for consumption. Since the seventeenth century, United Kingdom has been the leading European country in terms of herring production, ceding its position to Norway only between the two World Wars (Oras and Sammet 1982, 82; Schwach 2013).

For the Baltic area, the local subspecies of herring, *Clupea harengus membras*, is an important counterpart to the Atlantic herring. It lives east of the Danish Straights. Baltic herring is merely 20 centimeters long and its life lasts but for 6–7 years. The general name for Baltic herring processed for food in Estonian is *silk*; linguistic evidence reveals connections with Latvian and Lithuanian fish-eaters (Atlantic herring in Latvian is *silke* and in Lithuanian *silkė*). Atlantic herring is referred to by its international name, *heeringas*, in Estonian. According to Aliise Moora, the prominent scholar of Estonian food culture, the main food items for Estonian peasants have traditionally been black bread and Baltic herring. Atlantic herring gradually became an everyday food instead of a festivity food over the course of the nineteenth century (Moora 2007, 346, 368).

With the establishment of independence in 1918, the young Republic of Estonia faced numerous economic problems. One of the important tasks was to look for new export markets for domestic production. The vast market of the Russian Empire virtually disappeared for Estonian products behind the newly established state borders. Estonian foreign trade had to reorientate itself toward western Europe. The British government was among the first foreign countries to show interest in establishing a commercial treaty with the Republic of Estonia as they saw great potential in the transit trade through Estonia to its neighboring countries. The first official trade treaty between Estonia and the United Kingdom was signed in 1920. After that, the United Kingdom took second position after Germany in terms of Estonian foreign trade volume and became the most important export market for Estonian agricultural products until WWII (Pihlamägi 1999, 89–91). The trade and shipping treaty between the United Kingdom and Estonia stated, among other issues, that Estonia should promote employing British vessels, merchant and passenger ships alike (Pärna 1979, 87).

In 1929, 97% of the Atlantic herring imported to Estonia came from the United Kingdom (Pihlamägi 1999, 92). Smaller amounts were imported from other countries, such as Sweden and Norway. Estonian fishermen were engaged locally in coastal fishing, but the exploitation of ocean resources was well beyond their reach in the 1920s. Estonian periodicals monitored the statistics and published overviews of the fish trade on a regular basis. The comparison between the consumption of domestically produced Baltic herring and imported Atlantic herring was one of the central issues discussed (cf. *Päevaleht* [*Daily news*] 1923, 5).

The collapse of the international lending market after 1929 caused an impact on most of Central and Eastern European countries. The financial panic of 1931 drove those countries to restrict international payments by introducing exchange control measures (Irwin 1993, 90–119). The countries that had remained on the gold standard experienced overvaluation of their national currencies, which had a negative effect on their trade balance. In order to maintain a trade balance and preserve the higher value of their national currencies, states started using licensing systems, exchange control, clearing agreements, and other measures. The situation urgently raised the question of what goods Estonia should import and what could be produced domestically.

At that time, approximately 60% of the Estonian people earned their living from the agricultural sector (Valge 2003).[3] Therefore, it is not surprising that protectionist measures were implemented in Estonia especially with regard to agricultural produce. Foodstuffs and spices and drinks amounted to a third of total imports, but the figure dropped noticeably in 1930.

As for herring, in 1925 and 1926, its imports constituted 3.4% of the total value of Estonian imports, indicating its importance in the daily diet of the population of Estonia. Even in 1931, when the world market prices of agricultural produce and fish products had dropped, the figure was 1.3%, but in 1932 it only amounted to 0.3%. After 1932, the proportion of herring imports increased again and even reached the 1931 level at the end of the decade. However, it never returned to the level of the 1920s.

At the end of 1931, the Estonian *Riigikogu* [Parliament] passed the Organization of Goods Import Act that introduced the licensing system in Estonia (RT 1931, 90, 670). The Act stipulated that the government has the right to impose a national exclusive right of import with regard to certain goods, and it was regularly updated. The adoption of such an act was nothing new in the European context (Irwin 1993, 90–119). In essence, the Act gave the government the right to monopolize the import of certain goods. At the end of 1931, the government established control over nearly 40% of imports. The 1932 regulation included the import of herring in the licensing system, which gave the government the right to start regulating herring import quantities and thereby gain an additional mechanism for generating currency savings (RT 1932, 51, 450).

For the purpose of achieving currency savings, exchange control was implemented in Estonia in 1931. Under the conditions of exchange control, the government exercised control over the foreign currency received for the exported goods and decided how to divide it between the importers. The national bank obtained the exclusive right to perform foreign currency transactions. Similarly to the licensing system, the use of exchange control was common in Europe in that period. The largest country to use exchange control was Germany.[4]

The more successful functioning of exchange control was in turn facilitated by the licensing system. The fewer import permits the Ministry of Economic Affairs issued, the smaller the importers' demand for foreign currency from the Bank of Estonia was. A herring importer had to obtain an import permit from the Ministry of Economic Affairs. If currency savings were considered more important than herring imports, enterprises did not receive an import permit.

The licensing system with exchange control gave the state the right to regulate the quantity of imported goods. In the context of trade statistics, that meant a general reduction in foreign trade volumes. In 1932, foreign trade volume fell to 79.5 million kroons from the 132.3 million kroons of 1931, while only in 1929 it had amounted to 240.5 million kroons (see Figure 1).

Under the Organization of Goods Import Act, licenses were issued mainly for the goods that had no alternative in Estonia. As it was believed that domestic foodstuffs could easily substitute Atlantic herring, the import of herring was also decided to be licensed. The British government did not welcome this decision or the implementation of tariff-related measures.

The increase in tariff rates in several countries could be considered as an introductory step toward the world economy sinking into protectionism. In the summer of 1929, tariff rates were increased in Germany, France, and Italy and toward the end of the year also elsewhere in Europe. In 1931, several draft acts were presented to the Estonian *Riigikogu*, proposing radical increases in tariff rates that were considered as self-defense, creating a protective environment for the Estonian industry (ERA.969.3.270, 143; ERA.969.3.267, 13). The desire to increase tariff rates also entailed a monetary and fiscal policy objective.

According to initial plans, the customs duty on herring was to be tripled. The former minimum tariff[5] of 0.022 kroons per gross kilogram was to be replaced with 0.07 kroons (ERA.969.3.267, 143). It was presumed that the increase of the tariff rate would reduce the import of herring and help achieve currency savings. Many also

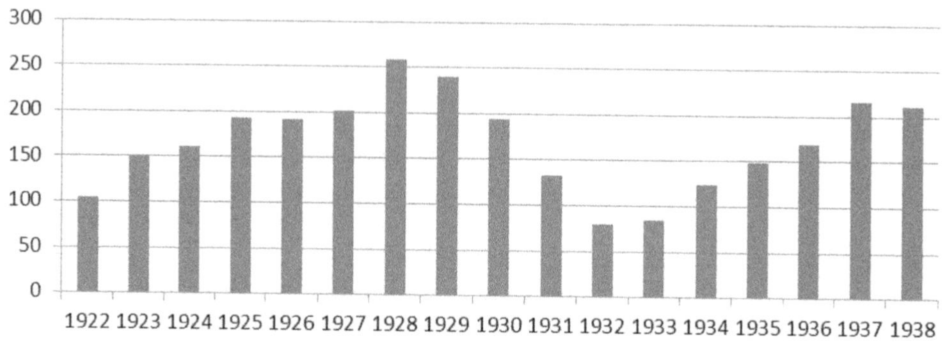

FIGURE 1 Estonian foreign trade volume, in millions of Estonian kroons.

Notes: The decrease in Estonian foreign trade volume clearly shows that the decline in the import of herring characterizes the whole of foreign trade. Foreign trade volumes decreased considerably more than GDP. By estimation, the world GDP decreased by 20% and the foreign trade volume by 40% during the Great Depression (Irwin 1993, 90–119). The downturn in foreign trade was largely caused by protectionism, which had started spreading intensively at the end of the 1920s.

Source: Pihlamägi (2004).

hoped that the higher tariff rate would help replace foreign herring with local Baltic herring and sprat, which were difficult to market elsewhere (ERA.969.3.267, 129). When the Act Amending the Basic Customs Tariffs Act was finally passed, the minimum tariff on herring remained unchanged at 0.022 kroons per gross kilogram (RT 1933, 63, 491). The regulation contains a notice: “Atlantic herring caught by Estonian ships outside the territorial waters of Estonia will not be taxed.” As it was considered important by legislators to include such notice in the text of the act, it evidently had caused some controversies and was thus regarded as an important issue to be clarified. It was stated in the explanatory memorandum that despite the relatively large proportion of the import of herring a considerable increase in the customs rate was not considered possible so as not to impair the people’s nutrition conditions (ERA.969.3.267, 496).

The Estonian Quest for Herring

Limiting herring imports by means of the licensing system resulted in a remarkable decrease in that particular segment of trade in 1932. A number of people active in the fishery and fishing business were discontent with both the expenses on imports and the prospective removal of salted Atlantic herring from the domestic market, which had for a long time been a traditional part of the Estonian food culture. The state campaign for replacing imported herring with domestically produced food yielded some rather biased opinion stories in local periodicals. Herring was explicitly constructed as the main competitor of “domestic” fish (A.G. 1930, 50)[6] and depicted as a tasteless “stomach filler” (Keller 1928, 237–39). It was pointed out that in Finland educated people preferred domestic fish to low-quality Atlantic herring that, in contrast, was favored in Estonian parvenu circles (A.H. 1928, 219).[7] It was even suggested that eating salted herring was a Russian custom characteristic of a "backward culture" (Šoberg 1930, 190). In addition to promoting domestic fish, however, Šoberg (1930, 191) also proposes "domestication" of herring fishing by blocking herring imports with high customs tariffs and developing an Estonian herring flotilla, following the example set by Finnish neighbors.

On January 21, 1931, the *Postimees* [Postman] daily – actually delivering false information – stated in a reproaching tone that Estonia imported Atlantic herring for almost 3 million kroons each year.[8] The main message of the article was – domestic production should be favored. Estonian periodicals of the time closely monitored the developments in fish production and sales in Finland, Latvia, Russia, and Poland. In 1930–1931, a sequence of short notices was published in the journal *Laevandus ja Kalaasjandus* [Shipping and Fishery] that quite clearly illustrate the escalating excitement about "taming" the herrings of far-away waters: in the August/September issue of 1930 (*Laevandus ja Kalaasjandus* 1930, 176), it was reported that Finland had sent its herring flotilla consisting of three trawlers and one mother ship to the Atlantic Ocean on June 1, and that they had returned at the beginning of September with more than 10,000 barrels of herring that had been sold in Finland with great success. In the June issue of the next year (*Laevandus ja Kalaasjandus* 1931a, 118), a report about a Polish herring fishing expedition appeared: encouraged by the success of the Finns, Polish fish traders planned to buy four to six trawlers from the Netherlands and set off for their own quest for herring. The continuing success story of the Finns was reported in August/September of 1931 (*Laevandus ja Kalaasjandus* 1931b, 193) – that year, two herring expeditions produced more than 20,000 barrels of herring and the public expressed hopes that soon the whole Finnish herring demand would be covered by domestically produced herring only. "As Finland lacks people competent in herring fishing, both expeditions were led by hired Norwegian experts, but in the nearest future the Finnish fishers expect to be competent enough to catch herring without any foreign help," the report concludes (*Laevandus ja Kalaasjandus* 1931b, 193).

Further public appeals about the need to diminish the import of herring in the context of the global economic crisis at the expense of favoring locally produced fish products were published in *Laevandus ja Kalaasjandus* in 1932 (J.J., 7/8, 14–15[9]; s.n. 11/12, 61–62). It was stressed that "our own" or Estonian specialists in herring fishing should be educated and new opportunities should be created for them, as currently there was neither experience nor expertise in ocean fishing in Estonia. The desire for alimentary sovereignty was an idea that swept across Europe during the interwar period, with Italy being the most famous example (Helstosky 2004, 1–26). The situation led to a daring move, the establishment of an Estonian national ocean flotilla for herring fishing in 1932. It is interesting to point out that the media coverage presented economic and ideological arguments, but no scientific surveys regarding the estimations of herring populations in the northern Atlantic or the general concerns of ocean resource exploitation that were available at that time (Rozwadowski 2002) seem to have been considered in devising the project.

On February 9, 1932a, *Päevaleht* reported without mentioning any names that "some businessmen, mainly mariners, were planning to start fishing for Atlantic herring near Iceland this spring." Forty female fish salters were to be hired from Tallinn, and some Norwegian specialists were to supervise them. The initial plan also included a hired Junkers hydroplane from which the surfacing fish shoals could be spotted.[10]

For the purpose of advancing domestic herring production, OÜ Kalandus [Fishery Company] was established in Tallinn on April 2, 1932 (Luhaveer 1996, 134). It had 18 shareholders, most of them natives of Northern Estonian coastal villages, the region where many young men traditionally chose to earn their living in maritime affairs. This was also the region that was most actively engaged in the illegal spirit trade with Finland in 1919–1932 (Pullat 1993, 183–96). As Prohibition was abandoned in Finland in April 1932, the men and vessels that had previously been engaged in smuggling were ready for new employment opportunities. On March 17, 1932b, *Päevaleht* published a thorough report on the preparations of the first Estonian herring expedition, stating plainly that three former spirit ships were to be rebuilt as herring trawlers. The link between spirit smuggling and herring fishing through legalizing the accumulated capital is also repeated in historiography (Oras and Sammet 1982, 82). The rebuilt trawling vessels were accompanied by a mother ship, *Eestirand,* that was bought specially for that purpose from the United Kingdom.

The necessary coal and salt for the first Estonian herring expedition were also bought from the United Kingdom. Herring nets, winches, and other hauling equipment, boats, and barrels were to be obtained from Norway as well as the hired "foreign specialists" to lead the fishing process and instruct in salting. In addition to Estonian fish salters, some of the "herring maids" also came from Finland. *Päevaleht* (March 17, 1932b) stated that the expedition would prevent spending foreign currency on herring imports and therefore support the protectionist politics of the Estonian government. As we can say in hindsight, the support was not mutual. The establishment of the Estonian herring fleet was also believed to relieve the unemployment caused by the global economic depression.

In order to initiate ocean fishing, an ideological agenda was developed to support the hoped-for economic bonuses. The quest for herring was constructed as an enterprise of national importance in the media. For propaganda purposes, journalist Evald Tammlaan[11] and film operators A. Hirvonen and Zimmermann[12] were hired by the fishery entrepreneurs to accompany the ships and document the whole event. As a result, an hour-length film about herring fishing[13] and a 24-part travelogue of the expedition titled *Yankee Man's Herring Letters* were produced. The travelogue featured preparations and departure of the expedition, the fleet's travel to the fishing grounds, descriptions of the Norwegian and Icelandic ports and countryside, but most importantly, it documented the daily life and work on the board of the ships, the practicalities of herring fishing, and commented about the economic context of the whole enterprise. This comprises abundant, yet often overlooked materials about the background of the Estonian "quest for herring," as well as of the daily experience of the Estonian herring fishers of the 1930s.

No scientists were included in the crew of the first Estonian herring expedition. In 1933, ichthyologist Aleksander Määr from the University of Tartu joined the herring fleet. His study results have been introduced briefly in local media (*Päevaleht*, September 13, 1933a; October 21, 1933b), but it remains unclear whether his contributions made it to international ocean studies.

The rhetoric used in newspaper reports on the herring flotilla constructs the entrepreneurs and crews as "our Vikings,"[14] who shall venture into the unknown in the quest for the "ocean vagabonds" (i.e., herrings). The mother ship is portrayed as "the hugest ship ever that has sailed the seas under the blue-black-white flag" and the flotilla is depicted as "our small and swaying colony that brings the colors of the Estonian flag to the eternal waves of the vast open seas" (*Päevaleht*, May 29, 1932c). This corresponds to Helen M. Rozwadowski's (2001, 220) observation about the mid-nineteenth century: marine naturalists and other ocean explorers shared the political motive of demonstrating national power.

The mother ship was sent off from Tallinn with a festive mess that was featured in the photo reportage on the front page of *Päevaleht* on June 11, 1932d. The Norwegian instructors were reported to be content with the equipment of the ships; also the light and dry fish barrels of Finnish origin were praised. The fleet arrived at its destination near the Icelandic waters at the beginning of July; a week later news about the first catch, 750 barrels of herring, of the glorious "herring hunters" was reported in *Päevaleht* (July 10, 1932e), followed by the front-page report about the expected arrival of Estonian herring production in Tallinn at the beginning of August. "And those are not some meagre fish, but of the first rate, the so-called king herrings," an anonymous writer announced in *Päevaleht* (July 15, 1932f).

The same news item also raised the question of tariff rates. Hope was expressed that if it was possible to sell cheaper herring with reduced tariff rates, more people could afford buying it. This would lead to considerable improvement of the overall nutritional conditions of the Estonian people. Thence, economic interests were veiled with the rhetoric of social welfare and national health.

The first actual catch of herring is depicted by Tammlaan as a major adventure. As it must have been the first encounter with live Atlantic herring for most of the crew, the excited tone of the reportage is only natural. The first alarm sounds at six in the morning, but the shoals descend before the boats complete besiege. After a couple of hours the maneuver is repeated and the first herring are caught in the nets. Tammlaan writes,

> And all of a sudden, the sea by the ship comes alive. Greenish-blue backs swarm and drop glittering scale spangles. A Norwegian jams an oar straight down – it stands – sways – is erect. It is as if pure silver gleams in the deep blue water. Fat, plump bodies jump into the air. . . . Men run and heap up at the big reservoir: everybody wants to be the first one to touch the first catch. They sure are big, fat fish – makes one wonder – if one remembers the thin, salty skins that one has sometimes mistakenly bought for herring in Estonia. Here the fish are so soft that one may squeeze it through one's fingers upon seizing it, just like a well-ripened plum – so that the fat drips from between the fingers. (*Päevaleht*, July 26, 1932a, 6)

The images of silver color and softness of the fish recur throughout the text. Traditionally, the image of fat is associated with affluence. The excitement of the catch, combined with the promise of the prospective wealth, is combined with a sense of wonder with regard to the fish as a live shoal: the density of the caught fish in the trawling net, the movement and the plumpness of the bodies, the tactile sensations that the crew members had never experienced before. We also see the continuation of the disparaging rhetoric pointed at the presumably low quality of the imported herrings.

Herring Catches and International Pressure

The descriptions of the routine fishing activities in the *Yankee Man's Herring Letters* bear the mark of (partly unfair) competition among the foreign ships that had gathered in the neutral waters close to the Icelandic coast where the herring shoals moved about. Internationally, the ICES had identified problems with overfishing in the North Sea in 1930, and governments were urged to survey new, offshore fishing areas (Rozwadowski 2002, 88). In practice, national fleets were competing with each other rather keenly at the previously known offshore spawning areas. This took place within the boundaries of the internationally agreed fisheries management areas that were delineated based on statistical (and political), not biological logic (Hubbard 2013, 93–4).

The Finnish newspaper *Ajan Sana* had reported Estonian fishers spying on the Finnish trawlers via their radio connections (*Päevaleht*, July 30, 1932g). "The feelings of kindred nations are to be kept separate from the business," as Tammlaan puts it in a subtitle for his reportage (*Päevaleht*, August 6, 1932b). He eloquently describes the "multinational chase" after herring shoals, triggered as soon as any of the ca. 30 ships lingering at the sight of each other have spotted signs of a shoal. "No law prevents such action – it is a free sea," he states. The one who is able to move to the shoal faster is the "winner" and gets the fish. "The sea swarms with boats and nets," he writes. "Curse words in Finnish, Norwegian and Estonian sound over the waves. Ships whistle for the sign of warning; the men in boats call monotonously 'ho-hoi' as they haul the herring nets to the boats" (*Päevaleht*, August 6, 1932b).

From this description, it becomes evident that the endeavor of herring fishing was far from being a heroic national conquest; the reality of the everyday practice on the ocean resembled more of elbowing one's way to herring in a close race between the herring hunters of different nationalities – Finnish, Norwegian, Polish, and Icelandic. According to the reports (*Päevaleht*, July 15, 1932f; September 3, 1932c), Spanish and French herring fleets had already left by that time; the Scottish fleet was fishing somewhat more off the coast, and the Latvians arrived only at the beginning of August.

The more experienced herring catchers probably perceived the ships of the Estonian herring fleet that had had no previous claims in the waters of the North Atlantic as foreign intruders. However, the smaller amount of catch per ship evidently caused more worries to the "herring chasers" than overfishing. The situation probably also contributed to the later collapse of fish stocks, as experienced in Norway (Schwach 2013, 104). Michael Graham explicates the pattern in "the great law of fishing" in 1935, suggesting that fisheries with unrestricted access eventually become unprofitable (Rozwadowski 2002, 91; Hubbard 2013, 91). Regardless of the harsh conditions both in and around the ships, OÜ Kalandus' first herring season was an economic as well as emotional success.

In June 1933, however, the United Kingdom demanded that Estonia promptly remove all obstructions to the import of herring. This demand can be regarded in the context of the general British foreign trade policy that focused especially on these groups of trade articles that had suffered the most notable declines during the Depression (Pihlamägi 2004, 280–82). At the end of the 1920s, herring was the main export article of the United Kingdom to Estonia and Latvia (Rooth 1993, 193). By 1932, the value of fish imports from the United Kingdom had dropped from 2.6 million kroons in 1926 to 0.1 million kroons, that is, the Estonian market was virtually lost for the British herring (Pihlamägi 2004, 273). The dramatic loss in the value of British herring production in 1931–1935 was not comparable to the slight decline in fishing quantities during the same period (*Lecture Notes on the Herring* 1938, 8). As the import of herring to Estonia had diminished disproportionally, the United Kingdom wanted to restore the earlier market situation.

Using diplomatic and economic pressure from the United Kingdom, the two countries concluded the so-called "Herring Agreement" on July 15, 1933. The agreement, signed for 1 year, entered into force on 22 July, that is, the same date on which the herring was expected to be excluded from the list of goods subject to the licensing system. The agreement had an immediate effect on the import of herring. In 1933, the total value of herring imports from the United Kingdom increased to 325,183 kroons. In just a year, from 1932 to 1933, the proportion of Atlantic herring imports to Estonia more than doubled (see Figure 2).

Thus, the increase in the import of herring after 1932 does not necessarily indicate principal alteration of the Estonian trade policy, but rather the United Kingdom as the most important trading partner asserting its wishes. As the export of goods to the United Kingdom formed approximately a third of Estonia's total exports, it would have been difficult for Estonia not to sign the 1933 Herring Agreement. Without the initiative of the United Kingdom, the import of herring could have been considerably

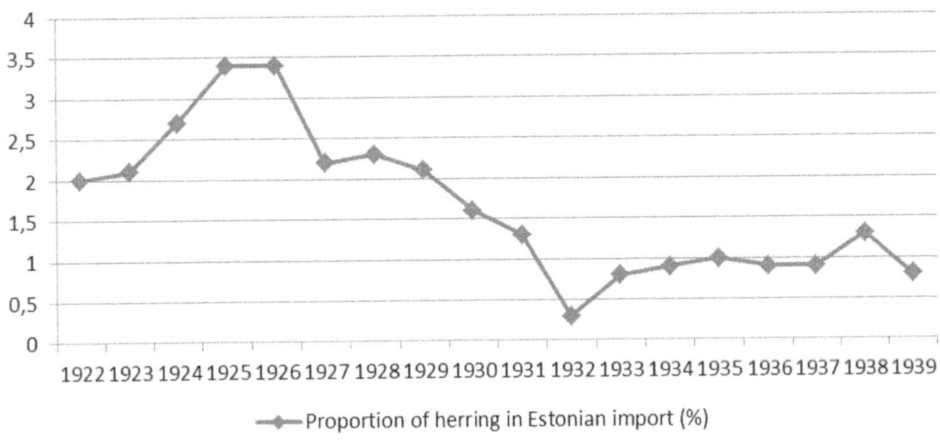

FIGURE 2 The percentage of herring in Estonian imports. The significant drop in herring imports in 1932 is clearly seen in this graph.

Source: Pihlamägi (2004).

more limited. Thence, the brave rhetoric of the national quest for herring in the local media was in reality replaced by the rationale of a foreign trade balance.

Conclusion

Food and its trade dynamics are influenced by both the ideological stance of the nation state and the international economic and political situation. The reasons for sudden leaps in the trade balance of traditional food items, such as Atlantic herring, cannot easily be explained away by changes in consumers' preferences. The reasons why Estonia experienced a sudden decrease in the import of Atlantic herring in 1932 were manifold. The Great Depression with the subsequent wave of protectionism in many countries, including Estonia, was the most evident reason. The Estonian government decided to apply an array of tariff- and nontariff trade measures to protect its domestic market, where foodstuffs, spices, and drinks formed a considerable segment. These measures turned out to have mixed results in the long term. The United Kingdom, as one of the major trade partners and export markets for Estonia, disagreed with a number of measures taken and applied diplomatic pressure to lift them. Licensing and raised tariffs played a central role in Estonia's import of Atlantic herring from the United Kingdom.

At the same time, local calls for national food sovereignty that was framed as the necessity to limit imports of foodstuffs that could be produced locally, on the one hand, consisted of the ideology-laden argument constructing Atlantic herring as an undesirable item in Estonian food culture. Following the same line of argumentation, an Estonian capital-based company, OÜ Kalandus, was established early in 1932, with the aim of founding Estonian ocean fishing using domestic capital and labor. The national media portrayed this first voyage of the fishing fleet as a heroic quest for Atlantic herring and a victorious conquering of distant waters and their riches.

Reality, however, differed from the ideal picture: the international waters off the coast of Iceland swarmed with the similar fleets of other European countries, and the Estonian workforce was inexperienced in encountering live herring. The domestic regulations concerning the tariff rates and the general foreign trade agreements of the Republic of Estonia of that time were not favorable to the enterprise either. As a result of the dynamics in international trade policies, Estonia's quest for Atlantic herring did not prove sustainable. Following the notes from the United Kingdom, it had to open its market to imports. This course of events did not solve the problem of overexploitation of the oceanic resources in the Atlantic Ocean, but the Estonian case study demonstrates how little such concerns weighed in the face of national and commercial interests.

Disclosure statement

No potential conflict of interest was reported by the authors.

Funding

This research was supported by the European Union through the European Regional Development Fund (Centre of Excellence CECT), by the Estonian Research Council grants Semiotic Modelling of Self-Description Mechanisms: Theory and Applications [IUT2-44], Dynamical Zoosemiotics and Animal Representations [7790], and EEA Norway Grants Animals in changing environments: Cultural mediation and semiotic analysis [EMP 151].

Notes

1. The retrospective analytical bibliography compiled in the Department of Bibliography in the Archival Library of the Estonian Literary Museum, and the portal of Digitized Estonian Newspapers, dea.nlib.ee, have been invaluably useful sources for obtaining material for the present article.
2. We wish to thank the State Archives of Estonia for providing assistance with the documents concerning the Estonian foreign trade regulations of the 1930s.
3. Agriculture included farming, animal husbandry, gardening, fisheries, and forestry.
4. Besides Germany and Estonia, Austria, Bulgaria, Czechoslovakia, Denmark, Greece, Hungary, Italy, Latvia, Lithuania, Poland, Portugal, Romania, Turkey, and Yugoslavia also implemented exchange control in Europe.
5. In the autumn of 1928, Estonia introduced two-tier customs tariffs: general customs tariffs and minimum customs tariffs. The goods of the countries that had concluded a trade agreement with Estonia were subject to the minimum customs tariff, while the products of countries that had not concluded an agreement with Estonia were subject to a set of general tariffs. The minimum customs tariff formed approximately 50% of the general customs tariff. For instance, at the beginning of 1931 the general customs tariff on herring was 0.045 kroons per gross kilogram and the minimum customs tariff 0.022 kroons per gross kilogram. It is more meaningful to focus on the minimum customs tariff, as Estonia had concluded agreements with nearly all of its important trade partners. In 1931, Lithuania, Spain, and Albania were the only European countries with which Estonia did not have a trade agreement.
6. Full name of the author is not registered in the Estonian Biographical Database.
7. Full name of the author is not registered in the Estonian Biographical Database.
8. According to archival sources, however, in 1931, the value of salted herring imported from the United Kingdom amounted to 782,392 kroons (ERA.1831.1.4349, 37) and in 1932 to 103,207 kroons (ERA.1831.1.4355, 34). The total value of the import of herring was approximately 0.8 million kroons in 1931 and approximately 0.1 million kroons in 1932 (Pihlamägi 2004, 216). The price drop that hit food products during the economic crisis naturally also reduced the value of imports, but the 1931 and 1932 herring prices did not considerably differ.
9. Full name of the author is not registered in the Estonian Biographical Database.
10. Four years earlier, a short notice had appeared in *Shipping and Fishery* about the "brilliant results" of using aeroplanes in herring fishing that enabled the behavior and location of the shoals to be monitored from the air and the information to be reported back to the flotilla (1928, 270). Estonian entrepreneurs later substituted the planned plane with "pioneer" motor boats as these were much less expensive and technically less demanding to operate.

11. According to the Estonian Biographical Database Tammlaan published under the pseudonym Jänkimees [Yankee Man], which was a general reference to seamen who had sailed across the Atlantic Ocean (cf. Past 1936).
12. Full names of the authors are not registered in the Estonian Biographical Database.
13. There is an anonymous news item about the festive screening of the film in Tallinn, published in *Päevaleht* on October 5, 1932h, but there is no record of such film in the Estonian Film Archives, nor are known the full names of the operators. There is a chronicle film dating to 1936 that includes less than 2 minutes of footage about the arrival of the herring fleet in the port of Tallinn. See filmi.arhiiv.ee/fis, search: 'heeringa' (accessed April 12, 2012).
14. The conceptualization of Estonians as vikings and Estonia as being part of the Scandinavian area already from times immemorial was a popular idea in the 1930s that was promoted in fiction and nonfiction alike (Kaljundi 2013).

References

Archival Sources

Eesti Riigiarhiiv (ERA)

Fond 969 Majandusministeerium

ERA.969.3.267, 13
ERA.969.3.267, 129
ERA.969.3.267, 143
ERA.969.3.267, 496
ERA.969.3.270, 143

Fond 1831 Riigi Statistika Keskbüroo

ERA.1831.1.4355, 34
ERA.1831.1.4349, 37

Printed Sources

A.G. 1930. "Väliskalakaubandus 1929." *Laevandus ja Kalaasjandus* 3 (46): 49–50.

A.G. 1931. "Kolme Miljoni Krooni Eest Heeringaid Aastas!" *Postimees*, January 21, 4.

A.H. 1928. "Märkeid Soome Kalaasjandusest." *Laevandus ja Kalaasjandus* 9 (29): 218–20.

Helstosky, C. F. 2004. "Fascist Food Politics: Mussolini's Policy of Alimentary Sovereignty." *Journal of Modern Italian Studies* 9 (1): 1–26. doi:10.1080/1354571042000179164.

Hubbard, J. 2013. "Mediating the North Atlantic Environment: Fisheries Biologists, Technology, and Marine Spaces." *Environmental History* 18 (1): 88–100. doi:10.1093/envhis/ems116.

Irwin, D. A. 1993. "Multilateral and Bilateral Trade Policies in the World Trading System: A Historical Perspective." *New Dimensions in Regional Integration*, edited by J. de Melo, and A. Panagariya, 90–119. Cambridge: Cambridge University Press.

Kaljundi, L. 2013. "Väljatung Kui Väljakutse. Eesti Viikingiromaanid ja Mälupoliitika 1930. Aastatel." *Keel ja Kirjandus* 8–9: 623–44.

Keller, A. 1928. "Kalakaupade Osa Meie Väliskaubanduses." *Laevandus ja Kalaasjandus* 10 (30): 237–39.

Klesment, M. 2000. "Eesti Väliskaubanduspoliitikast Kahe Maailmasõja Vahelisel Perioodil (Eesti-Saksa Kaubandussuhete Baasil)." *Ajalooline Ajakiri* 1: 81–88.

Luhaveer, O. ed. 1996. *Mereleksikon*. Tallinn: Eesti Entsüklopeediakirjastus.

Moora, A. 2007. *Eesti Talurahva Vanem Toit*. Tartu: Ilmamaa.

Oras, R., and J. Sammet. 1982. *Lahekäärust Ookeaniavarustele. Eesti Kalanduse Ajaloost*. Tallinn: Valgus.

Past, E. 1936. *Meresõidu Romantikat ja Traagikat*. Tallinn: Tallinna Merekooli Lõpetanud Kaugesõidkuaptenite Ühingu Kirjastus.

Pullat, R. 1993. "Aus der Geschichte des Schmuggelhandels mit Spiritus zwischen Estland und Finnland in den 1920-30er Jahren." *Scripta Mercaturae* 27 (1/2): 183–96.

Pärna, A. 1979. *Meri Ja Mehed. Meresõidust Eestis*. Tallinn: Valgus.

Pihlamägi, M. 2004. *Väikeriik Maailmaturul. Eesti Väliskaubandus 1918–1940*. Tallinn: Argo.

Pihlamägi, M. 1999. "Eesti Kaubandussuhted Suurbritanniaga Aastail 1918–1940." *Acta Historica Tallinnensia* 3: 88–108.

Põltsam, I. 2008. *Liivimaa Väikelinn Varase Uusaja Lävel. Uurimus Uus-Pärnu Ajaloost 16. Sajandi Esimesel Poolel*. Tallinn: TLÜ Kirjastus.

Raud, V. 1934/35. "Rahvusvahelise Kaubanduspoliitika Uusi Meetodeid." *Konjunktuur* 2: 162–68.

Rooth, T. 1993. *British Protectionism and the International Economy*. Cambridge: Cambridge University Press.

Rozwadowski, H. M. 2002. *The Sea Knows No Boundaries: A Century of Marine Science under ICES*. Seattle: University of Washington Press.

Rozwadowski, H. M. 2001. "Technology and Ocean-scape: Defining the Deep Sea in Mid-Nineteenth Century." *History and Technology*. 17: 217–47. doi:10.1080/07341510108581993.

Schwach, V. 2013. "The Sea around Norway: Science, Resource Management, and Environmental Concerns, 1860–1970." *Environmental History* 18 (1): 101–110. doi:10.1093/envhis/ems107.

Sicking, L., and D. Abreu-Ferreira, eds. 2009. *Beyond the Catch: Fisheries of the North Atlantic, the North Sea and the Baltic, 900–1850*. Leiden: Brill.

Šoberg, J. 1930. "Soolaheeringate Sisseveo-tollide Tõstmine on Põhjendatud." *Laevandus ja Kalaasjandus*. 10: 190–191.

Valge, J. 2003. "Uue Majanduse Lätteil. Eesti Sisemajanduse Kogutoodang Aastatel 1923–1938." *Akadeemia* 10–12: 2202–28, 2443–87, 2712–35.

Periodicals

Laevandus ja Kalaasjandus

J.J. 1932. "Meie Valitsuse Imelikud Sammud Eesti Ulgumere Kalastuse Arendamisel." *Laevandus ja Kalaasjandus* 7/8: 14–15.

s.n. 1928. "Heeringapüük Õhulaevade Kaasabil." *Laevandus ja Kalaasjandus* 11: 270.

s.n. 1930. "Soome Atlandi Heeringapüük Õnnestus." *Laevandus ja Kalaasjandus* 8/9: 176.

s.n. 1931a. "Poolakad Lähevad Heeringa Püügile." *Laevandus ja Kalaasjandus* 6: 118.

s.n. 1931b. "Soomlaste Heeringalaevastik Kasvab." *Laevandus ja Kalaasjandus* 8/9: 193.

s.n. 1932. "Veel Heeringast ja Heeringatollist." *Laevandus ja Kalaasjandus* 11/12: 61–62.

Päevaleht

Jänkimees [Tammlaan, E.] 1932a. "Kuidas Püüti Eesti Esimesed Heeringad." *Päevaleht*, July 26, 6.

Jänkimees [Tammlaan, E.] 1932b. "Jänkimehe Heeringakirjad." *Päevaleht*, August 6, 6.

Jänkimees [Tammlaan, E.] 1932c. "Jänkimehe Heeringakirjad." *Päevaleht*, September 3, 5.

s.n. 1923. "Silkude, Heeringate ja Kilude Sisse- ja Väljavedu ja Tarvitamine Eestis." *Päevaleht* October 26, 5.

s.n. 1932a. "Eesti Laev Heeringapüügile." *Päevaleht*, February 9, 4.

s.n. 1932b. "Piirituselaevad Heeringaid Püüdma." *Päevaleht*, March 17, 3.

s.n. 1932c. "20 000 Heeringatündrit Välismaalt, Mida Oleks Võidud Kodumaal Teha." *Päevaleht*, May 29, 7.

s.n. 1932d. "Heeringapüügi Emalaev "Eestirand" Asus Teele." *Päevaleht*, June 11, 1.

s.n. 1932e. "Eesti Heeringalaevastiku Rikkalik Saak." *Päevaleht*, July 10, 3.

s.n. 1932f. "Värsked Heeringad Tulevad Turule." *Päevaleht*, July 15, 1.

s.n. 1932g. "Eesti Heeringalaevad Loevad Soome 'Salakeeli.'" *Päevaleht*, July 30, 2.

s.n. 1932h. "Film Eesti Heeringapüügist." *Päevaleht*, October 5, 5.

s.n. 1933a. "Islandist Tagasi Teaduslike Kollektsioonidega." *Päevaleht*, September 13, 1.

s.n. 1933b. "Kaugvete Kalastus Eesti Tuleviku Ala." *Päevaleht*, October 21, 6.

s.n. 1938. *Lecture Notes on the Herring*. London: Herring Industry Board Publications.

Riigi Teataja, official gazette for the publication of Estonian laws

RT 1931, vol. 90, art 670.

RT 1932, vol. 51, art 450.

RT 1933, vol. 63, art 491.

Index

Note: **Boldface** page numbers refer to figures and tables, page number followed by n denote endnotes.